RURAL BUILDING

VOLUME TWO

RURAL BUILDING

VOLUME TWO

Basic Knowledge

INTERMEDIATE TECHNOLOGY PUBLICATIONS 1995

Published by Intermediate Technology Publications Ltd,
103–105 Southampton Row, London WC1B 4HH, UK
Under licence from TOOL, Sarphatistraat 650, 1018 AV, Amsterdam, Holland

© TOOL

A CIP catalogue record for this book
is available from the British Library

ISBN 1 85339 315 0

Printed in Great Britain by SRP, Exeter, UK

PREFACE

This official text book is designed purposely to meet the needs of trainees who are pursuing rural building courses in various training centres administered by the National Vocational Training Institute.

The main aim of this book is to provide much needed trade information in simple language and with illustrations suited to the understanding of the average trainee.

It is the outcome of many years of experiment conducted by the Catholic F.I.C. brothers of the Netherlands, and the German Volunteer Service instructors, in simple building techniques required for a rural community.

The National Vocational Training Institute is very grateful to Brothers John v. Winden and Marcel de Keijzer of F.I.C. and Messrs. Fritz Hohnerlein and Wolfram Pforte for their devoted service in preparing the necessary materials for the book; we are also grateful to the German Volunteer Service and the German Foundation For International Development (DSE) - AUT, who sponsored the publication of this book.

We are confident that the book will be of immense value to the instructors and trainees in our training centres.

DIRECTOR: National Vocational Training
Institute, Accra

© Copyright
by Stichting Kongregatie F.I.C.
Brusselsestraat 38
6211 PG Maastricht
Nederland

Alle Rechte vorbehalten
All rights reserved.

INTRODUCTION TO A RURAL BUILDING COURSE

Vocational training in Rural Building started in the Nandom Practical Vocational Centre in 1970. Since then this training has developed into an official four year course with a programme emphasis on realistic vocational training.

At the end of 1972 the Rural Building Course was officially recognised by the National Vocational Training Institute. This institute guides and controls all the vocational training in Ghana, supervises the development of crafts, and sets the examinations that are taken at the end of the training periods.

The Rural Building programme combines carpentry and masonry, especially the techniques required for constructing housing and building sanitary and washing facilities, and storage facilities. The course is adapted to suit conditions in the rural areas and will be useful to those interested in rural development, and to farmers and agricultural workers.

While following this course, the instructor should try to foster in the trainee a sense of pride in his traditional way of building and design which is influenced by customs, climate and belief. The trainee should also be aware of the requirements of modern society, the links between the old and new techniques, between traditional and modern designs -- and how best to strike a happy medium between the two with regard to considerations like health protection, storage space, sewage and the water supply. The trainee should be encouraged to judge situations in the light of his own knowledge gained from the course, and to find his own solutions to problems; that is why this course does not provide fixed solutions but rather gives basic technical information. The instructor can adapt the course to the particular situation with which he and the trainee are faced.

This course is the result of many years of work and experimentation with different techniques. The text has been frequently revised to serve all those interested in Rural Development, and it is hoped that this course will be used in many vocational centres and communities. It is also the sincere wish of the founders of this course that the trainees should feel at the completion of their training that they are able to contribute personally to the development of the rural areas, which is of such vital importance to any other general development.

We are grateful to the Brothers F.I.C., the National Vocational Training Institute and the German Volunteer Service for their assistance and support during the preparation of this course.

Bro. John v. Winden (F.I.C.)
Wolfram Pforte (G.V.S.)
Fritz Hohnerlein (G.V.S.)

LAY-OUT OF THE RURAL BUILDING COURSE

The Rural Building Course is a block-release-system course, which means that the trainee will be trained in turn at the vocational centre and at the building site. The period of training at the centre is called "off-the-job" training, and the period on the building site is called "on-the-job" training. Each will last for two years, so that the whole course will take four years and will end with the final test for the National Craftsmanship Certificate.

BLOCK RELEASE SYSTEM

YEAR	TERM 1	TERM 2	TERM 3
1	X	X	X
2	O	O	O
3	O	X	O
4	X	O	X

X = OFF-THE-JOB TRAINING
O = ON-THE-JOB TRAINING

The total "off-the-job" training period is approximately 76 weeks, each week 35 hours. During this training about 80% of the time is spent on practical training in the workshop. The remaining 20% of the time is devoted to theoretical instruction.

The total "on-the-job" training period is approximately 95 weeks, each week 40 hours. During this period the trainee does full-time practical work related to his course work. In addition some "homework" is assigned by the centre and checked by the instructors.

A set of books has been prepared as an aid to the theoretical training:
- A - Rural Building, Basic Knowledge (Form 1)
- B - Rural Building, Construction (Forms 2, 3, 4)
- C - Rural Building, Drawing Book (Forms 1, 2, 3, 4)
- D - Rural Building, Reference Book

All these books are related to each other and should be used together. The whole set covers the syllabus for Rural Building and will be used in the preparation for the Grade II, Grade I, and the National Craftsmanship Certificate in Rural Building.

CONTENTS

BOOK INTRODUCTION

SAFETY FIRST

PART 1: BASIC MASONRY TECHNIQUES — 2

TECHNICAL TERMS — 2

BONDING OF WALLS — 6
- General effect of bond / Half-block bond / Other types of bond

HANDLING THE TROWEL — 10
- Preparing the bed joint / Removing surplus mortar

THE FIRST WALL — 13
- Setting out / Organizing the work / Preventive measures / The first blocks / The first course / The second course / Toothed end / Stopped end / Marking and cutting blocks

QUOINS — 26
- Second course / Walling between quoins / Fixing the mason line / Use of the tingle plate / Completing the course

BONDING PROBLEMS — 32

CORNER BONDS — 32
- Corner bond for footings / Corner bond for a rising sandcrete wall / Corner bond for rising landcrete wall

T-JUNCTION BONDS — 38
- T-junction bond for footings / T-junction bond for rising sandcrete walls / T-junction bond for rising landcrete walls

CROSS JUNCTION BONDS — 46
- Cross junction bond for footings / Cross junction bond for rising sandcrete walls / Cross junction bond for rising landcrete walls / Footings - Plinth course - Rising landcrete wall

PIERS — 60
- Attached piers / Simple piers / Footings for piers / Bonds for sandcrete piers / Bonds for landcrete piers

PART 2: BASIC CARPENTRY TECHNIQUES — 72

TECHNICAL TERMS — 72

TESTING BOARDS	74

- Winding strips / How to test small work pieces / How to test long work pieces

MEASURING AND MARKING OUT TIMBER	78

- Measuring with a rule / Marking out with a rule / Marking with a pencil / Marking with a chalk line

TIMBER CONSTRUCTIONS	81
PREPARATION OF TIMBER	82

- Sequence of operations for preparing timber / Timber marks / Marking of frames / Cutting list

FASTENING WITH NAILS	92

- Driving nails / Holding power / How to prevent splitting during nailing / Finishing off

FASTENING WITH SCREWS	96

- Holding power / Driving screws in soft wood / Driving screws in hard wood

ANGLE JOINTS	98

- Nailed butt joint / Plain mitred joint / Rebated butt joint / Sequence of operations for constructing a box with this joint / Housed joint / Sequence of operations for constructing the joint / Common mortice and tenon joint for box-like constructions / Sequence of operations for constructing the joint / Cornerlocked joint / Sequence of operations for constructing the joint

FRAMING JOINTS	112

- Halved joints / Sequence of operations for constructing the joint / Corner-halved joint / Cross-halved joint / Stopped tee-halved joint / Common mortice and tenon joint for frame-like constructions / Sequence of operations for constructing the joint / Haunched mortice and tenon joint / Stub tenon joint / Twin tenon joint / Securing the joints / Bridle joint / Sequence of operations for constructing the joint / Corner bridle joint

WIDENING JOINTS	130

- Plain glued butt joint / Dowelled widening joint / Rebated joint / How to plane a rebate with an ordinary rebate plane / How to plane a rebate with an adjustable rebate plane / Loose tongued joint / How to plane a groove for a loose tongued joint

MISCELLANEOUS CARPENTRY TECHNIQUES 138
- Marking a board to fit an irregular surface / Measuring the width of openings / Marking out irregular designs with templates

PART 3: PREPARATION FOR ON-THE-JOB TRAINING 140

BUILDING PRELIMINARIES 140
- Plan / Plot and site clearing / Site organization

SETTING OUT 142
- 3-4-5 method / Lining out / Direct marking / Using the plumb bob to mark the foundations / Using the large square / Using the mason square

FOUNDATIONS 148
- Excavating the foundation trenches / Marking the depth of the concrete and levelling the trench / Foundation concrete / Casting - Compacting - Levelling / Curing

FOOTINGS 150
- Purpose of footings / Height of footings

HARDCORE FILLING 152
- Function of the hardcore filling / Methods of filling and compaction

PLINTH COURSE 154
- Function of the plinth course

OPENINGS 156
- Door openings / Window openings

SCAFFOLDING 158
- Working scaffolds / Ladders

BRIDGING OPENINGS 162
- Methods of bridging

LINTELS 164
- Built-up wooden lintel / Reinforced concrete lintel / Formwork for a reinforced concrete lintel / Casting reinforced concrete lintels

ROOFS 172
- Anchor beam / Wall plate / Terms

PLASTER - RENDER — 174

- Function of the plaster or render / Number of coats / Plastering or rendering / The aggregates / Mix proportions / Additional protective measures / Spatterdash

FLOOR CONSTRUCTION — 179

- One-course work / Two-course work / Casting methods Shrinkage gaps / Expansion gaps

A WORD FOR THE TRAINEE BEFORE ON-THE-JOB TRAINING — 184

BOOK INTRODUCTION

Rural Building, Basic Knowledge is your first construction book. This book plus most of the Reference Book (Rural Building Tools, Maintenance of Tools, Materials and Building Products) will be treated in the first year of centre training.

This book is divided into three parts:

PART 1: BASIC MASONRY TECHNIQUES

This part covers the very basic techniques of masonry. These include the preparation of mortar, blocklaying, the proper arrangement of blocks, and building up walls. Some of the techniques mentioned in the section on the arrangement of blocks are techniques used mainly for bricklaying and therefore only apply to areas where bricks are available.

PART 2: BASIC CARPENTRY TECHNIQUES

The basic techniques covered here include planning construction pieces, preparing wood for construction pieces, ways of fastening the pieces and the important types of joints and how to construct them.

PART 3: PREPARATION FOR ON-THE-JOB TRAINING

This part is meant to be a preparation for the trainee's first year of on-the-job training. It should enable him to follow the basic procedures he is confronted with on a building site, and to understand the technical terms used there. This part of the book doesn't attempt to give detailed information about technical problems but merely to give an idea of them and to enable the trainee to understand the terms and deal with situations on the building site. Most of the procedures will be treated more intensively in the Construction and Drawing books.

At the beginning of both the carpentry and masonry sections there is a list of the terms you will need to know, together with explanations.

You will often be asked to refer to one of the supplementary books for additional information. Especially in the first part of the course, much of the basic knowledge you will need about tools and materials and products will be found in the other books.

You should prepare yourself for each lesson by reading the material before class and looking up the references given in the text for the Reference Book or Drawing Book. If you are not familiar with a tool that is mentioned, now is the time to look in the Reference Book and learn about the tool and how to use it.

There is ample space provided in the book for making notes and sketches.

SAFETY FIRST

When you first start doing construction work it is essential to realize the importance of being safety conscious. You must develop safe working habits in order to prevent injuries to yourself and others.

Accidents can generally be avoided by using ordinary care and skill. Most accidents are a result of thoughtlessness or carelessness on the part of some person.

To practise any kind of craftwork you need your hands, your legs and feet, a healthy body, and most of all your head.

Safety first means that you use your head and think out what you are going to do before going ahead with the job. By first thinking the operation through, you will discover that there is a correct way of doing the task, and some other ways of doing it that may cause danger to yourself and others. Look in the beginning of your Reference Book for a list of general safety rules.

Follow the safety rules, but also use your own sense. When you realize that certain actions can be dangerous, you can plan to prevent accidents and injuries. Look ahead to find the dangerous points of a task and plan to make them safe by taking proper precautions. We can make hundreds of safety rules, but they are useless unless we understand why they are needed and we all cooperate in following them.

One of the most important safety precautions is learning to use the right tool for the job, and in the correct way. The correct way is the safest way.

RESPECT OTHER PEOPLE: RESPECT YOURSELF!

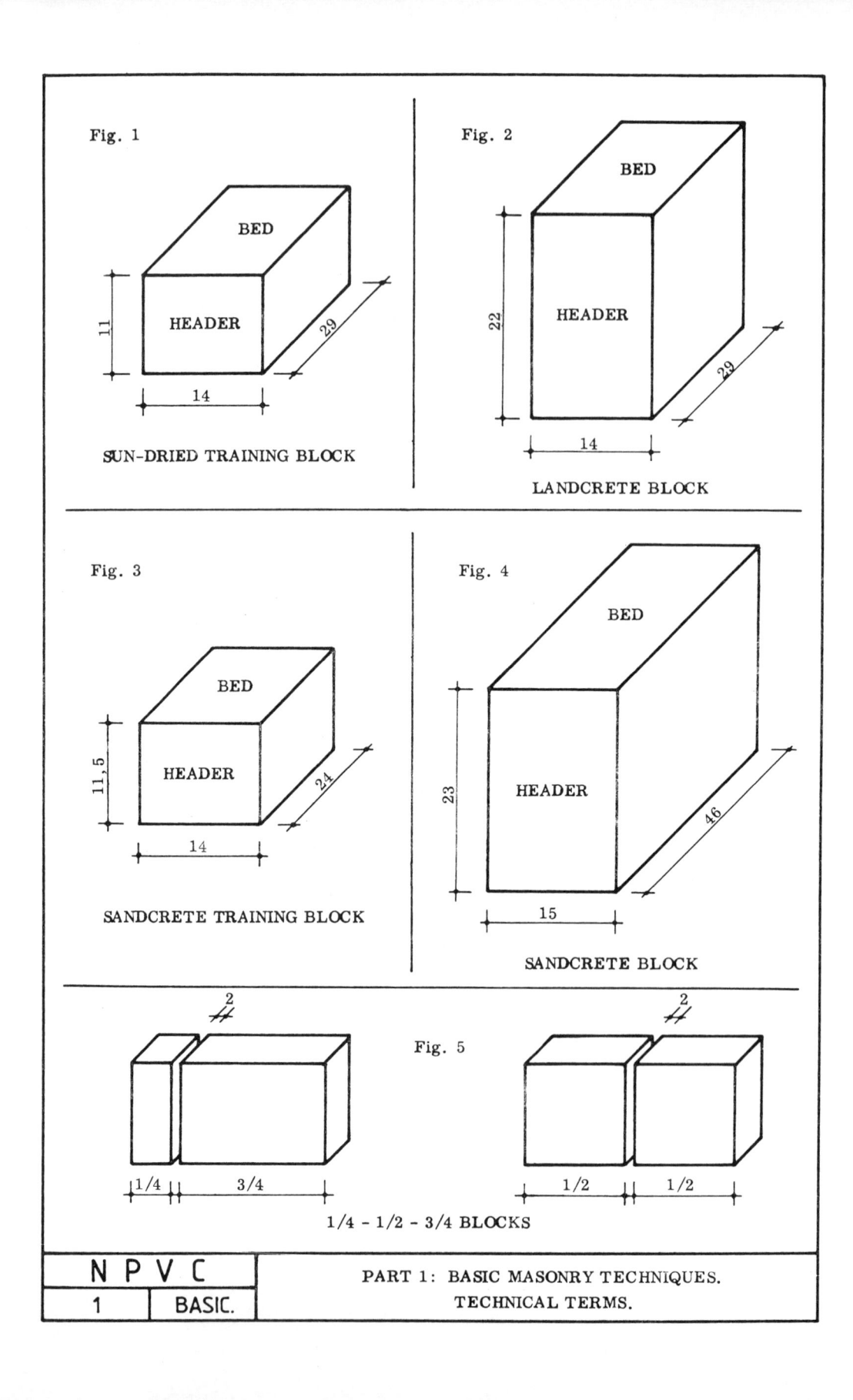

Fig. 1
BED
HEADER
11
14
29
SUN-DRIED TRAINING BLOCK

Fig. 2
BED
HEADER
22
14
29
LANDCRETE BLOCK

Fig. 3
BED
HEADER
11,5
14
24
SANDCRETE TRAINING BLOCK

Fig. 4
BED
HEADER
23
15
46
SANDCRETE BLOCK

Fig. 5
2
1/4   3/4
2
1/2   1/2
1/4 - 1/2 - 3/4 BLOCKS

N P V C	PART 1: BASIC MASONRY TECHNIQUES.
1 BASIC.	TECHNICAL TERMS.

PART 1: BASIC MASONRY TECHNIQUES

TECHNICAL TERMS

Before describing the methods used in bonding it is necessary to briefly define and explain a few of the technical terms commonly used.

- SUN-DRIED TRAINING BLOCKS: These blocks, as the name implies, are dried in the sun, because they do not contain cement (Fig. 1). They are often used by beginners during their first terms of in-centre training, as they are easy to make and to handle.

    The dimensions of these blocks are approximately 29 cm x 14 cm x 11 cm, which allows us to construct complicated bonds with them. They can be used to make an excellent inside wall also, as the following chapters will describe.

- LANDCRETE BLOCKS: These are blocks used for actual building (Fig. 2).

- SANDCRETE TRAINING BLOCKS: These are specially made small sandcrete blocks with the approximate dimensions of 24 cm x 14 cm x 11,5 cm; so they can be used to teach the making of more complicated bonds (Fig. 3).

    Like the sun-dried blocks, they are often used in training because they are easy to handle. As they are made with cement they can be used for a long time before they wear out.

- SANDCRETE BLOCKS: This type of block is used for actual building (Fig. 4).

- 3/4 - 1/2 - 1/4 - BLOCKS: These are parts of blocks obtained by cutting a block through either the centre line or the quarter line, less half the thickness of the joint (Fig. 5). The cut is made along the width, not along the length. A special block gauge may be used for marking off the different sizes.

- HEADER: A block is known as a header when it is placed in a wall so that its smallest face is exposed (Fig. 1, next page).

- BED: The under-surface of a block, or the mortar on which the block is laid.

- STRETCHER: If the biggest face of a block is exposed, the block is called a stretcher (Fig. 1, next page). This is the way most blocks are laid in Rural Building, and we say that the block is laid edgewise. If a block is laid flatwise, so it is actually showing the top face, it is also called a stretcher (Fig. 1, next page).

- COURSE: This is the term applied to each layer or row of blocks, with the bed joint included (Fig. 1, next page).

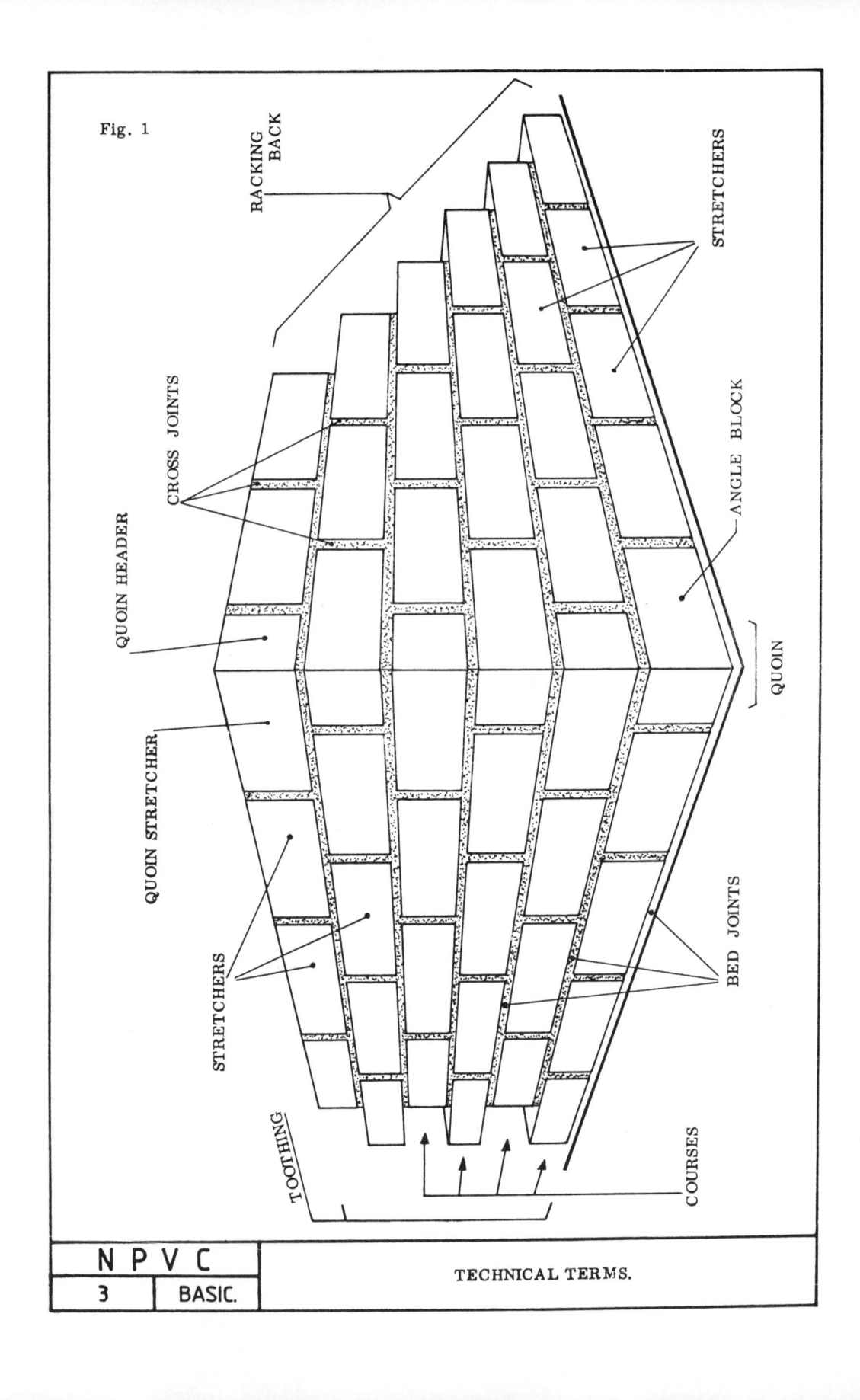

Fig. 1

TECHNICAL TERMS.

NPVC — 3 — BASIC.

- BED JOINT: This is the horizontal mortar joint between two courses (Fig. 1).

- CROSS JOINT: The vertical joints between the blocks (Fig. 1).

- QUOIN: The quoin is the outside corner of a wall or the external angle on the face side of the wall (Fig. 1).

- ANGLE BLOCK: This is the block which actually forms the corner in each course (Fig. 1).

- STOPPED END: A plain vertical surface which forms the end of a wall (see page 21).

- TOOTHED END OR TOOTHING: The form produced at the end of a wall by recessing every other course by half a block (Fig. 1) in order that the wall may be extended later using the same bond.

- RACKING BACK: As an alternative to toothing, the end of a wall may be set back half a block at each course (Fig. 1). This is also done so that the wall may be extended later using the same bond.

- BUILDING UNIT: This refers to the dimensions of a full block, plus one joint. For sandcrete blocks the building unit is 25 cm high by 48 cm long (2 cm joints).

- FOUNDATION: The base, usually concrete, on which the building rests. It is usually set below ground level, and is the only part of the building in direct contact with the ground.

- FOOTINGS: The courses laid directly on top of the foundations; usually three flatwise courses of sandcrete blocks (see page 35).

- PLINTH COURSE: The edgewise course of sandcrete blocks laid on top of the footings (see page 35).

- RISING WALL: The edgewise courses of sandcrete or landcrete blocks which build up the rest of the wall (see page 35).

NOTES:

HEAVY LOAD

Fig. 1

Fig. 2

HEAVY LOAD

Fig. 3

POLE GIVES WAY
AND SINKS DOWN

LARGER AREA OF BOARD
PREVENTS POLE FROM
SINKING DOWN

LOAD

Fig. 4

LOAD

Fig. 5

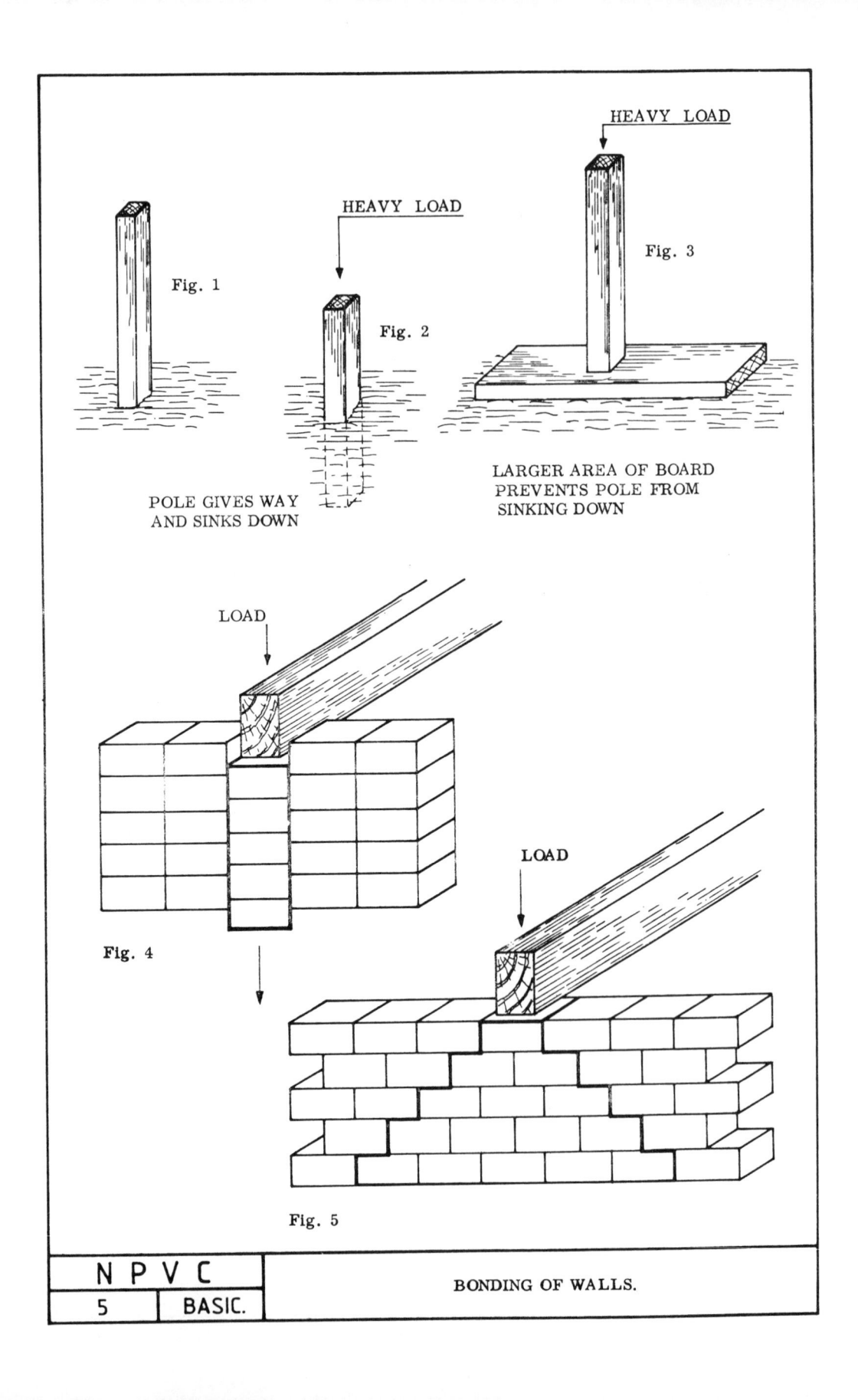

N P V C	
5	BASIC.

BONDING OF WALLS.

BONDING OF WALLS

The practise of blocklaying requires a complete understanding of the correct arrangement of the blocks forming a wall. This correct arrangement of blocks, regardless of the method, is known as bonding.

The blocks are placed so that they overlap each other and care must be taken to ensure that as far as possible no vertical joint is immediately above another vertical joint in the course below.

GENERAL EFFECT OF BOND

Fig. 1 shows a wooden pole placed vertically on soft ground. If this pole has to carry a heavy load it will sink down into the soil, because the total area on which the pole rests on the ground is far too small to support it (Fig. 2).

A possible solution to the problem is shown in Fig. 3: a board is laid flat on the ground and it now carries the pole with the load. In this way the total load is distributed over a larger area of ground and it is impossible for the pole to sink down.

If a wall is built up by simply placing blocks directly above each other, we say the wall is built without bond. If a heavy load is put on top of this wall (Fig. 4) the column of blocks immediately under the load tends to give way and sink down.

In order to make the wall stronger in itself and able to distribute loads properly, the Rural Builder applies the so-called half-block bond.

A properly bonded wall which receives a heavy load will distribute the pressure over a large number of blocks and therefore over a much greater area (Fig. 5).

NOTES:

Fig. 1   HALF-BLOCK BONDING (LANDCRETE)
SCALE 1:10 (cm)

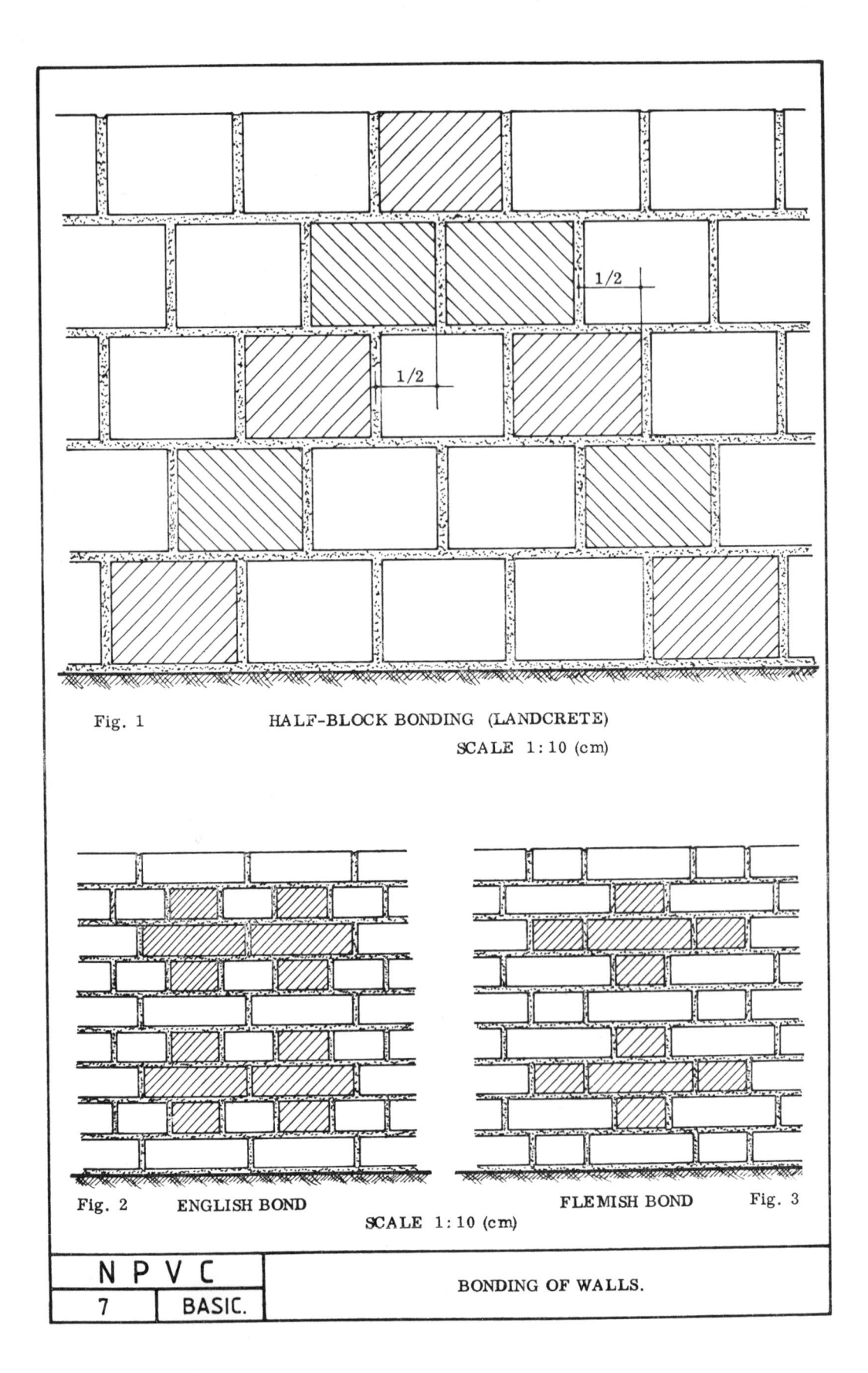

Fig. 2   ENGLISH BOND   FLEMISH BOND   Fig. 3
SCALE 1:10 (cm)

N P V C	BONDING OF WALLS.
7 BASIC.	

HALF-BLOCK BOND

The simplest form of bonding is that where all the blocks are laid down as stretchers, each block overlapping the one below by half its length (Fig. 1).

This form of bonding is only suitable where a one-block thickness of the wall is sufficient. In Rural Building the most common wall thickness is 14 cm when landcrete blocks are used and 15 cm when using sandcrete blocks; provided that they are laid edgewise.

If for some reason a thicker wall is required, the blocks may be laid flatwise. By doing this the wall thickness will be increased to 23 cm, 29 cm or even 46 cm depending on the type of block and the way the blocks are laid.

OTHER TYPES OF BOND

There are many types of bond in use, the two most common being the English Bond (Fig. 2) and the Flemish Bond (Fig. 3), both of which are used with bricks.

Bricks are smaller blocks with approximate dimensions of 24 cm x 11,5 cm x 7 cm.

- ENGLISH BOND: This bond consists of alternate courses of headers and stretchers. The centre of any stretcher is in line with the centre of the header in the courses above and below.

- FLEMISH BOND: This bond consists of alternate headers and stretchers in the same course. Again, the centre of any stretcher is in line with the centre of the header in the courses above and below.

NOTES:

BONDING OF WALLS.

N P V C
BASIC.   8

Fig. 1

CORRECT GRIP

MOVE

Fig. 2

3 cm thick

Fig. 3

Fig. 4

Fig. 5

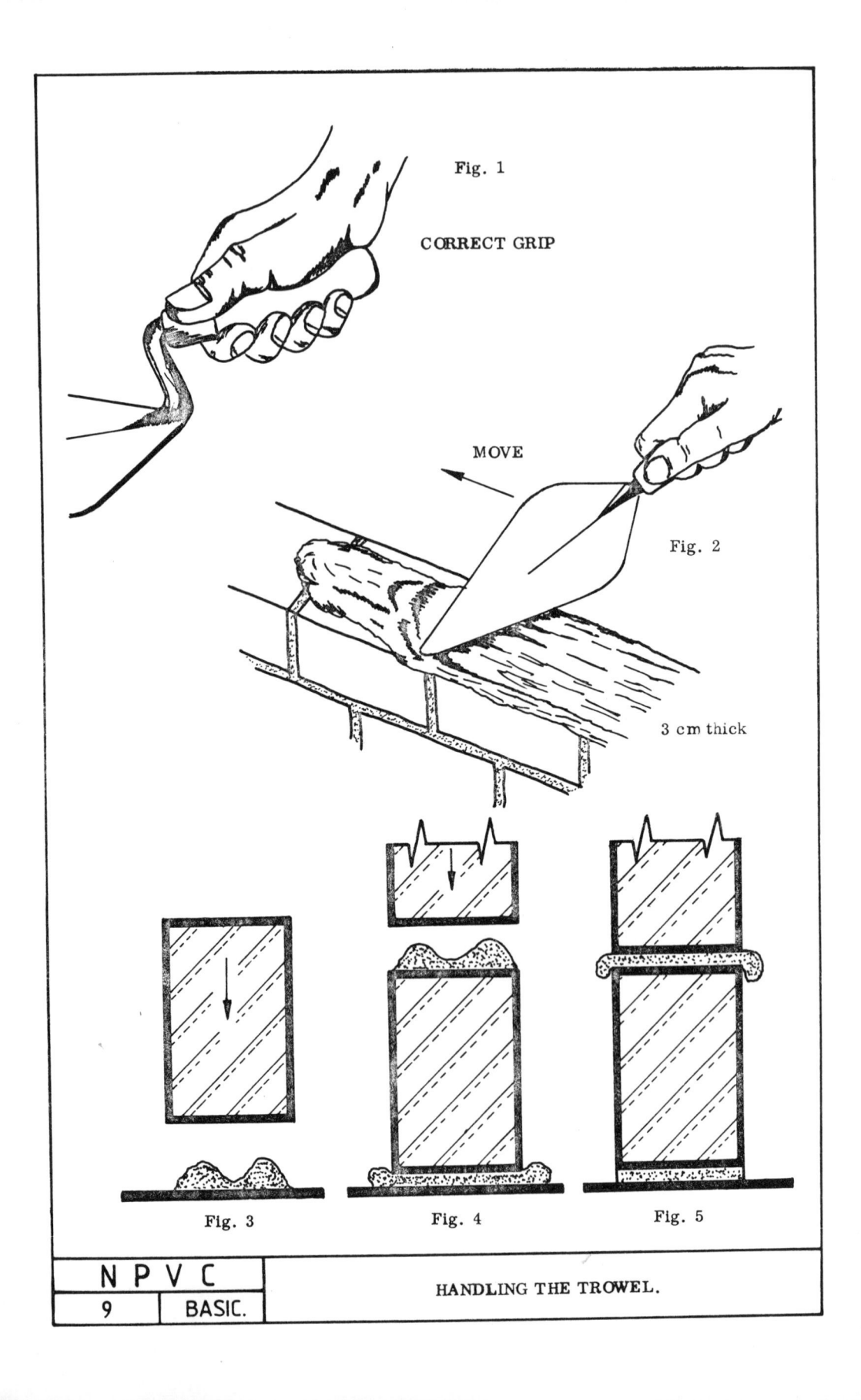

N P V C		HANDLING THE TROWEL.
9	BASIC.	

HANDLING THE TROWEL

In the process of laying blocks, the brick trowel is used to perform a series of operations during which the trowel is seldom put down or changed from one hand to the other.

All operations require free and easy manipulation of the trowel from the wrist and it is therefore essential to master the correct handling of the trowel.

Fig. 1 illustrates the correct grip on the handle, with the thumb resting on the ferrule. The thumb must be in this position in order to manipulate the trowel skillfully.

The amount of mortar picked up from the headpan will depend on the nature of the job, but for the trainee it is advisable to pick up a sufficient amount to lay one training block, i.e. a heaped trowel. Later, when working with the common big blocks, it will become necessary to pick up two trowelfuls of mortar.

PREPARING THE BED JOINT

Place the mortar in the middle of the wall or the marked position of the first course and spread it out by a pushing movement with the back of your trowel, into a layer about 3 cm thick (Fig. 2).

Next draw the point of your trowel through the centre of the layer, making a mortar bed suitable for the block (Fig. 3).

A block laid on a bed prepared in this way will at first rest on the two outer edges, but when it is pressed down to its correct position it will not only squeeze mortar from the front and back of the block but will also squeeze it into the valley in the centre. In this way a solid bed is formed (Figs. 3, 4, & 5).

- NOTE: All the tools mentioned in this section on masonry techniques are described in the Rural Building Reference Book, pages 3 to 15.

NOTES:

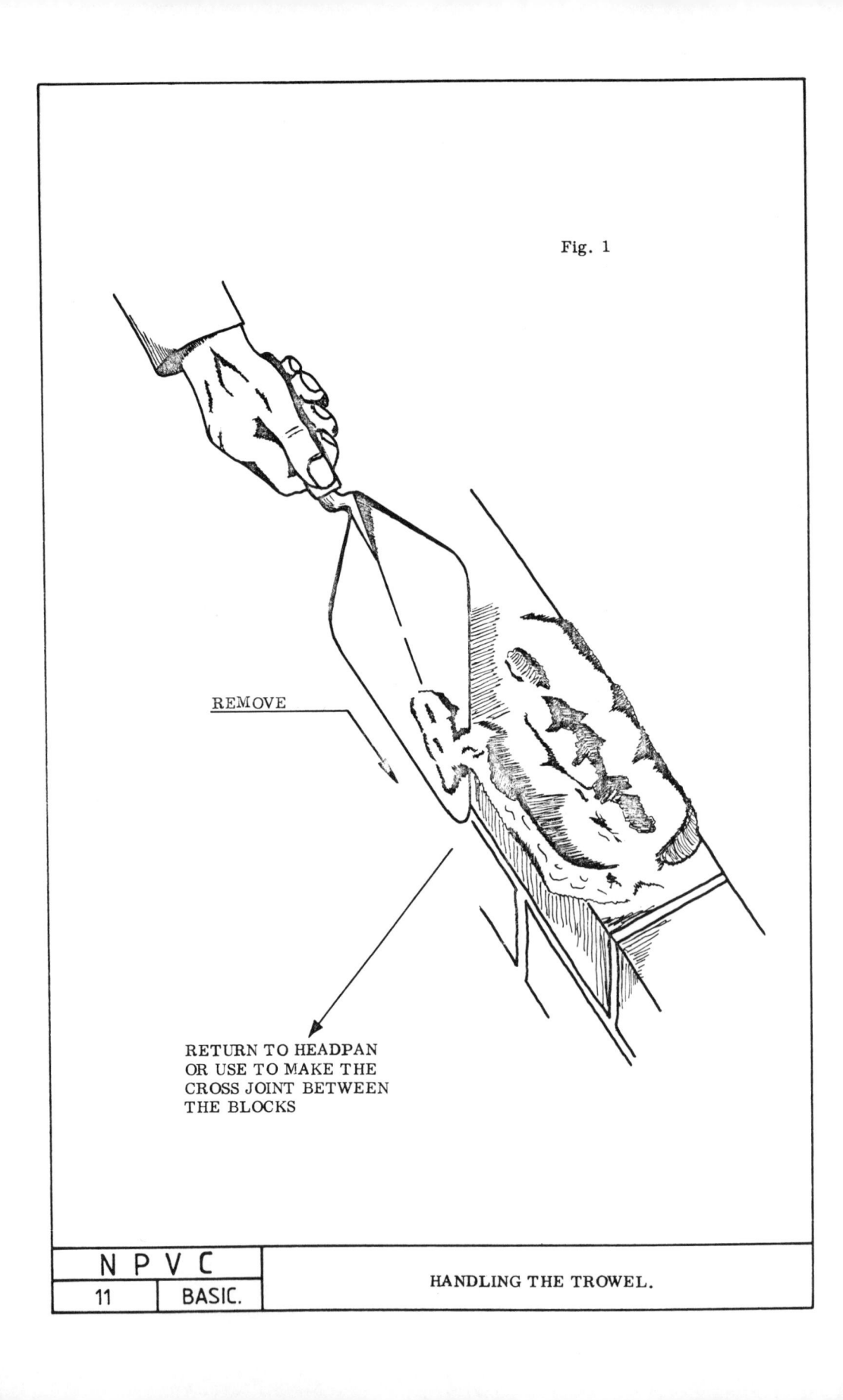

Fig. 1

REMOVE

RETURN TO HEADPAN
OR USE TO MAKE THE
CROSS JOINT BETWEEN
THE BLOCKS

N P V C	
11	BASIC.

HANDLING THE TROWEL.

REMOVING SURPLUS MORTAR

Before and after the block is layed, a certain amount of mortar will project from both sides of the wall. This must be removed before it drops down, as one of the most important principles of the Rural Builder is to work as economically as possible. This means saving materials. Fig. 1 shows the position of the trowel for these operations.

The surplus mortar recovered on the trowel is usually taken to form the cross joint between the last laid block and the previous one, or it is returned to the headpan.

NOTES:

THE FIRST WALL

SETTING OUT

Before actual building operations are started, you must know the correct position and dimensions of all the parts of the building.

This information is given in the plan or drawing of the building, which of course must have been already prepared and at hand.

The positions of the walls, for instance, have to be marked on the ground according to the measurements given on the plan before any building operation starts.

This operation is called "setting out" and we will deal with it repeatedly here, because it is one of the most important preparatory steps in building.

ORGANIZING THE WORK

The workplace has to be well organized in order to operate smoothly and safely.

Building materials such as blocks and mortar should be neither too close nor too far away from the wall being erected. A working space of about 90 cm will usually be all right. The blocks should be neatly stacked, not just thrown in a heap; and there should always be an adequate supply available, so that work is not delayed by waits for materials.

Keep your tools together and near your workplace so they are within easy reach. When you use a tool, put it back immediately afterwards so that it cannot fall off the wall etc. and injure you or other workers. Make a habit of putting your tools down in a way that prevents accidents.

- NOTE: You cannot expect to produce a good job with your tools and materials always scattered around. Neatness and orderliness show the professional.

  Never throw, kick or drop tools as you might damage them.

  Work on one side of the wall only. As the wall becomes higher, you won't be able to move from side to side anyway.

NOTES:

PREVENTIVE MEASURES

Almost all of the building in the Northern and Upper Regions of Ghana is done during the dry season, with its high temperatures and low humidity. These conditions are important and our building procedures must take them into consideration to prevent problems with drying out.

Before you put the mortar down and spread it, thoroughly wet the top of the foundation or the already laid course.

Do the same thing with the block that you are going to lay next. This is to prevent the block from absorbing too much moisture from the mortar. The porous landcrete or sandcrete blocks quickly suck in any moisture they come into contact with. This process is known as absorption. If the blocks absorb too much moisture from the mortar, it will not be able to set properly, and the joints will be weak. By sprinkling sufficient water onto the blocks, we ensure that there will be enough moisture left in the mortar to allow it to harden properly.

It is also important to never spread too much mortar at one time. Some masons prepare the mortar bed in advance for five, six or even more blocks in order to speed up the work. This is wrong.

While the first blocks are placed, lined-out, and levelled, the rest of the mortar is exposed for too long to the sun and air.

Due to the high temperatures and the low humidity, the mortar dries out very fast. As a result the mortar becomes too stiff, making it difficult to lay the last blocks and weakening the grip between the mortar and the block. The end result is a weak wall.

The Rural Builder should always keep in mind the dry climate and never spread more mortar than is actually needed.

NOTES:

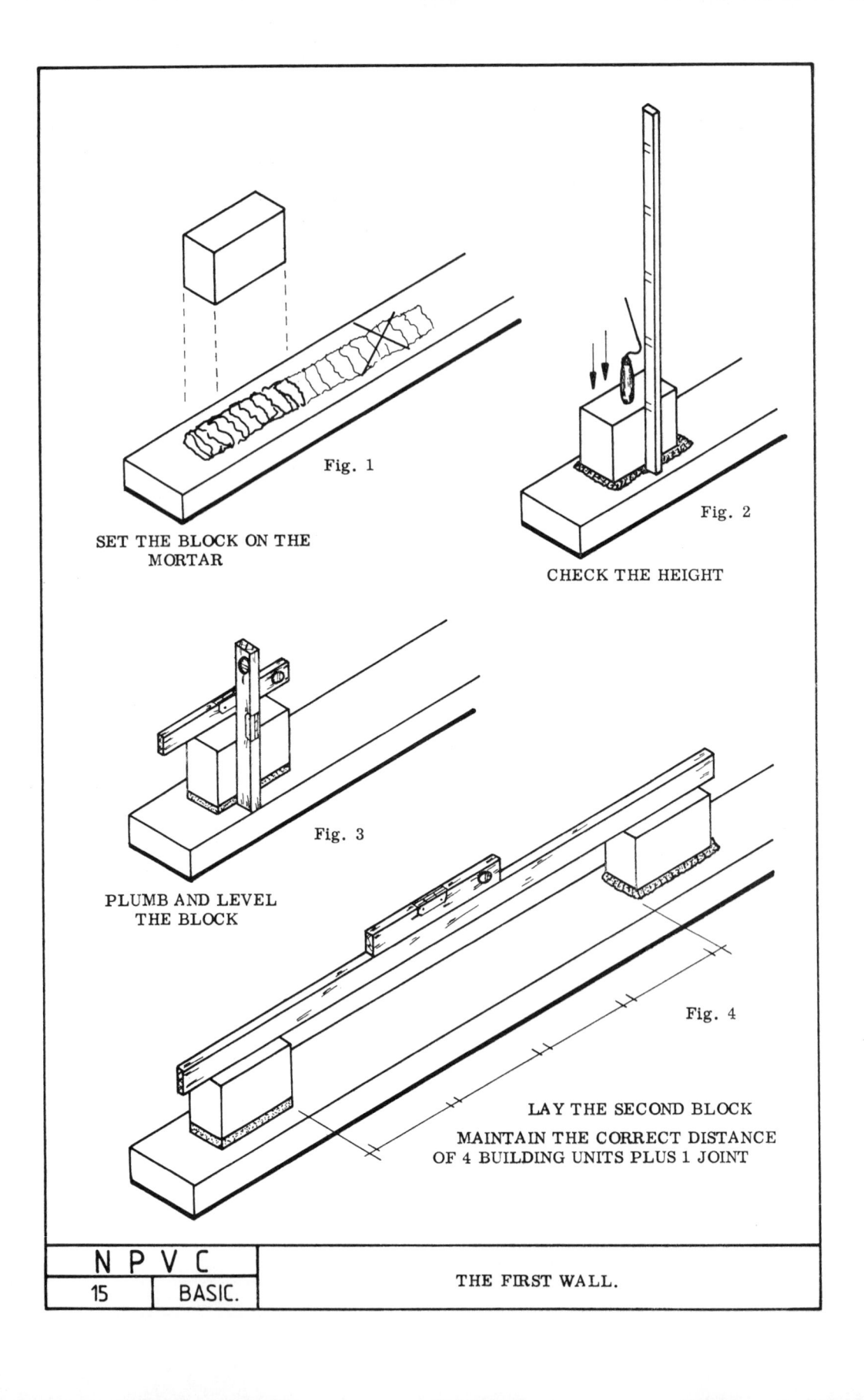

Fig. 1

SET THE BLOCK ON THE MORTAR

Fig. 2

CHECK THE HEIGHT

Fig. 3

PLUMB AND LEVEL THE BLOCK

Fig. 4

LAY THE SECOND BLOCK

MAINTAIN THE CORRECT DISTANCE OF 4 BUILDING UNITS PLUS 1 JOINT

N P V C	
15	BASIC.

THE FIRST WALL.

THE FIRST BLOCKS

After you have wetted the block and the area where it is supposed to be set, spread the mortar according to the method described on page 12. Set the block immediately onto the mortar bed and press it down firmly and evenly (Fig. 1).

If the bed has been spread correctly, only a few taps with the handle of the trowel will be needed to adjust the height of the block. The height is checked by comparing the height of the block with the gauge marks on the straight edge (Fig. 2).

Next, plumb the block with the spirit level along the stretcher face and the header face as shown in Fig. 3. The pressure on the block will have squeezed out some of the mortar. Trim off this excess, collect it on the trowel and return it to the headpan.

If you don't use enough water to make the bed, the block will not come up to the required height; it will sit too low. If on the other hand too much mortar is used, the block will sit too high.

Do not try to correct problems like this by pushing some mortar from the edges towards the inside of the bed using your fingers; or by knocking hard on the top of the block to try and force it down. These are very poor practices. Instead, simply remove the block and re-spread the mortar.

At the beginning of the training you will have to re-lay blocks quite often. As you gradually gain experience you will be able to spread just enough mortar to lay one block, without any of the problems mentioned above.

Lay the second block at a distance of four building units and one joint away from the first block (Fig. 4). Hold the straight edge against the stretcher faces of the two blocks to make sure that they are in line.

NOTES:

Fig. 1

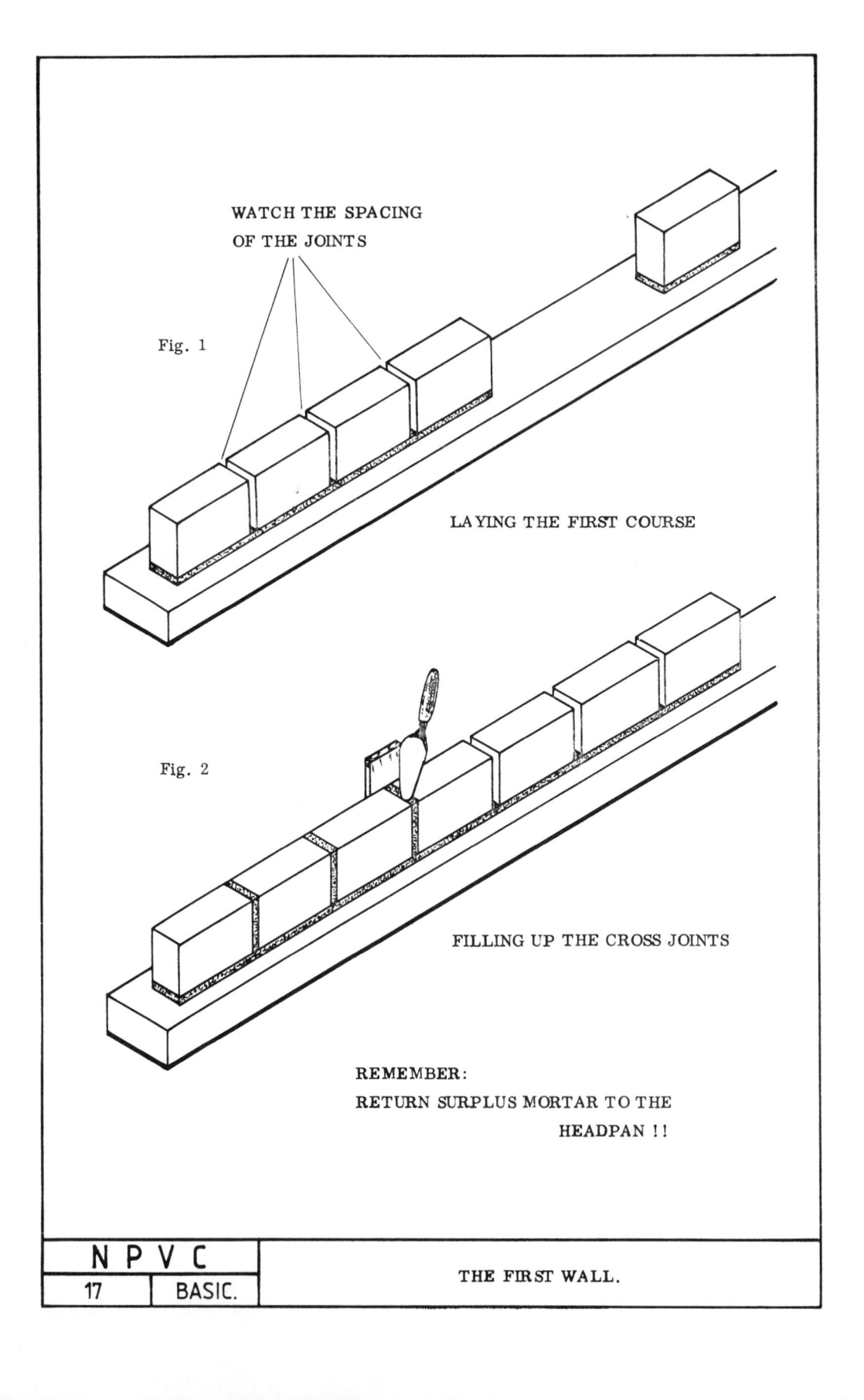

WATCH THE SPACING
OF THE JOINTS

LAYING THE FIRST COURSE

Fig. 2

FILLING UP THE CROSS JOINTS

REMEMBER:
RETURN SURPLUS MORTAR TO THE
HEADPAN !!

N P V C	
17	BASIC.

THE FIRST WALL.

THE FIRST COURSE

Because the first two blocks are in line and at the same height, we can complete the course without using the spirit level, only using the straight edge. Starting from either block (but still working on only one side of the wall) more blocks are inserted between the first two (Fig. 1).

Their height is adjusted by placing the straight edge on top and pressing the blocks down until the top surfaces of all the blocks touch the straight edge equally, along their whole length.

Line out the course (make it perfectly straight) by holding the straight edge against the stretcher faces and moving the blocks until they touch it along their full length.

During these operations take care to maintain the proper distances between the blocks. The next step is to fill the remaining open gaps between the blocks with mortar, thus forming the cross joints. This job is done by closing the back of the gap with the aid of a small wood float while carefully pushing the mortar down into the joint with the trowel (Fig. 2).

All cross joints must be completely filled up with mortar so that no holes are left, which would reduce the strength of the course and the whole wall.

All the excess mortar which has dropped down or was squeezed out of the bed must now be collected and returned to the headpan to be mixed with the rest of the mortar.

NOTES:

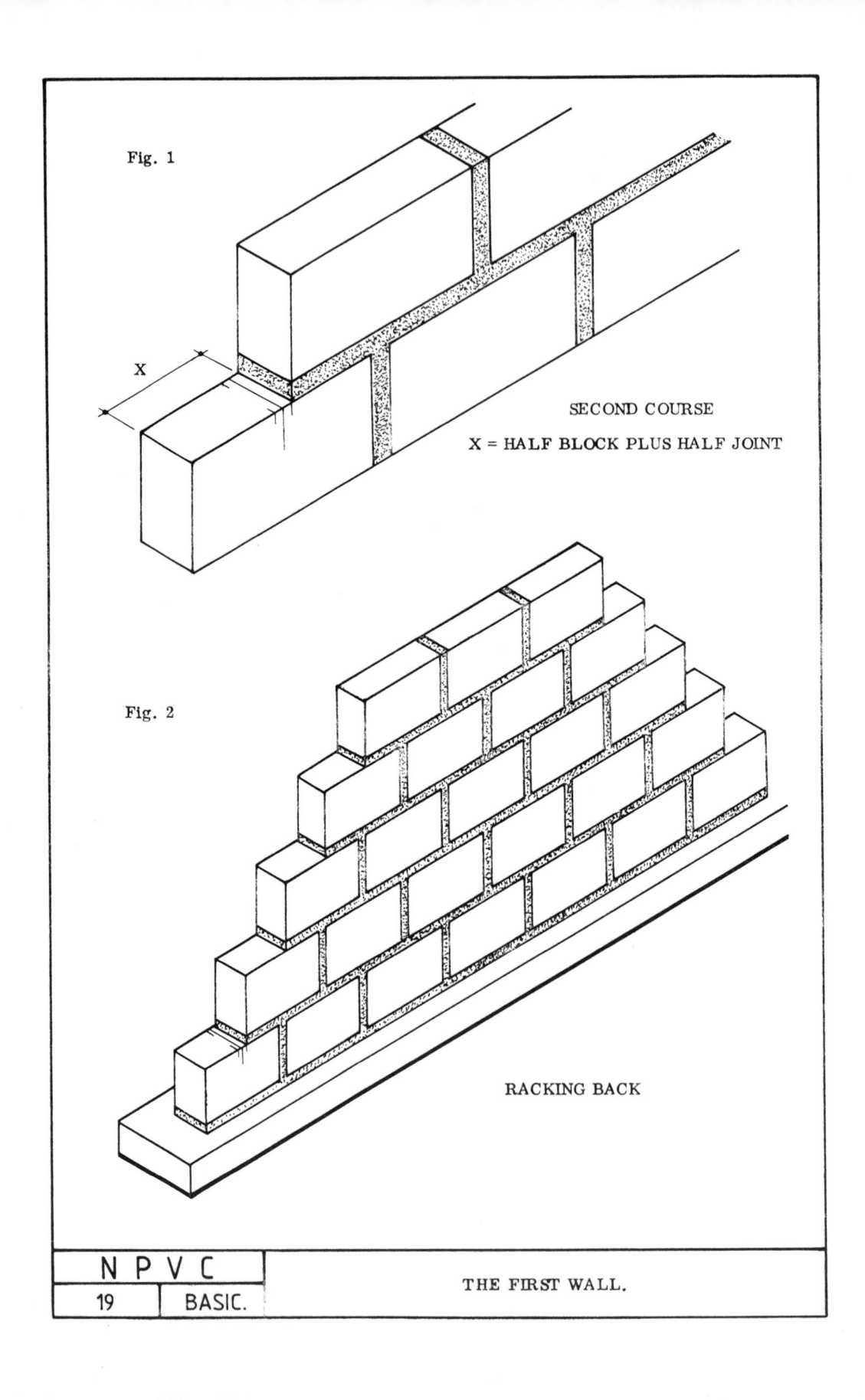

Fig. 1

SECOND COURSE

X = HALF BLOCK PLUS HALF JOINT

Fig. 2

RACKING BACK

N P V C		
19	BASIC.	THE FIRST WALL.

THE SECOND COURSE

Lay the first block of the second course with its centre exactly above the first cross joint so that it overlaps both blocks below equally.

No matter what sort of wall-ending is desired, the first block of the second course is always a full block laid above the first cross joint between two stretchers. This is known as the 1 - 2 - 1 rule. By doing this you maintain the half-block bond throughout the wall (Fig. 1)

After you check the height of this block, you must plumb its face. Hold the spirit level vertically along the face of the lower block with one hand, while with the other hand you move the upper block until its face is also in full contact with the spirit level, and the bubble is in the centre of the tube.

Follow the same operation with the second block, laying it above the last cross joint in the lower course. Insert the remaining blocks between them according to the method used for the first course.

The construction of any subsequent course is merely a repetition of the above operations and will result in a wall with racking back at both ends (Fig. 2).

- NOTE: If the Rural Builder has a choice between racking back and toothing, racking back should be the preferred method. This is because the joints used in toothing are difficult to fill properly when completing the wall, which often results in a weak grip all along the joints.

NOTES:

Fig. 1

TOOTHED END

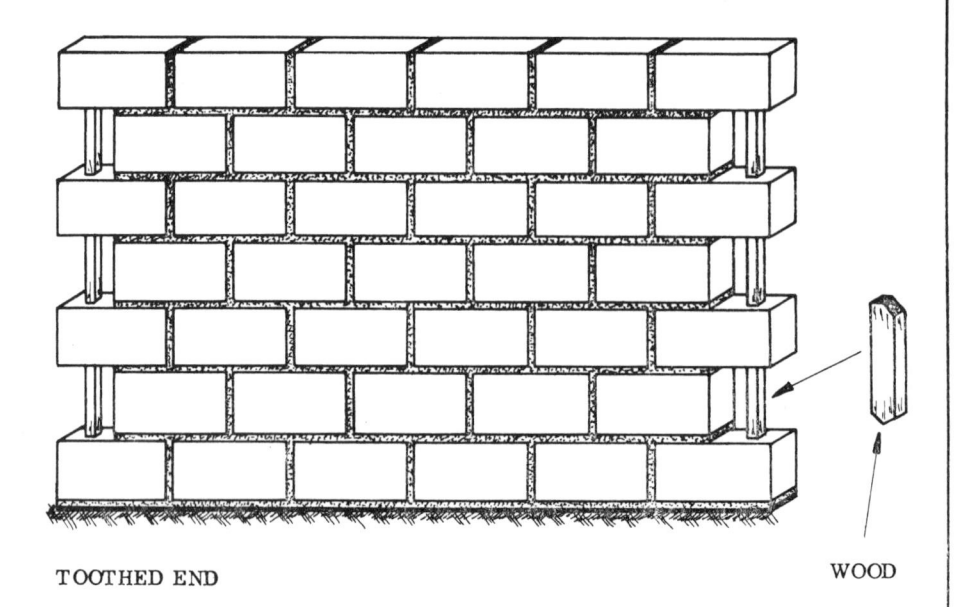

WOOD

Fig. 2

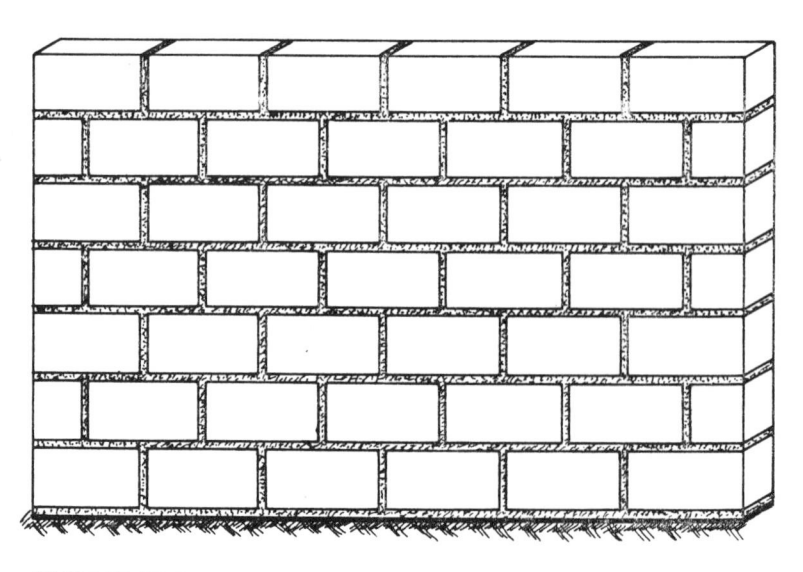

STOPPED END

N P V C	
21	BASIC.

THE FIRST WALL.

TOOTHED END

The construction of a wall with a toothed end starts with the same operations used for a wall with a racking back. After you lay the first block of the third course, the next block you lay forms the toothed end. As this block projects past the one below by more than half its own length, it should normally tip over on its projecting end. To prevent this, a temporary support must be provided until the block is overlapped by the first block of the fourth course and the mortar has set hard. This temporary support is preferably a short piece of board cut to the height of the block plus two bed joints. The upper and lower ends of the supporting board may be chamfered slightly to keep it from wedging between the blocks when it is removed (Fig. 1).

This block must be exactly above the first block of the first course. To ensure this, hold the spirit level against both header faces and make any necessary corrections. It is not necessary to level the top face if the stretcher and header faces have been plumbed.

The construction of further courses is again only a repetition of these operations.

STOPPED END

The construction of a wall with a stopped end is very similar to that with a toothed end. The sequence of operations is exactly the same except that instead of the supporting board, half a block is added at the end of the second course before the third course is laid (Fig. 2).

This means that there are no gaps left as in the toothed end, so it is called a stopped end.

NOTES:

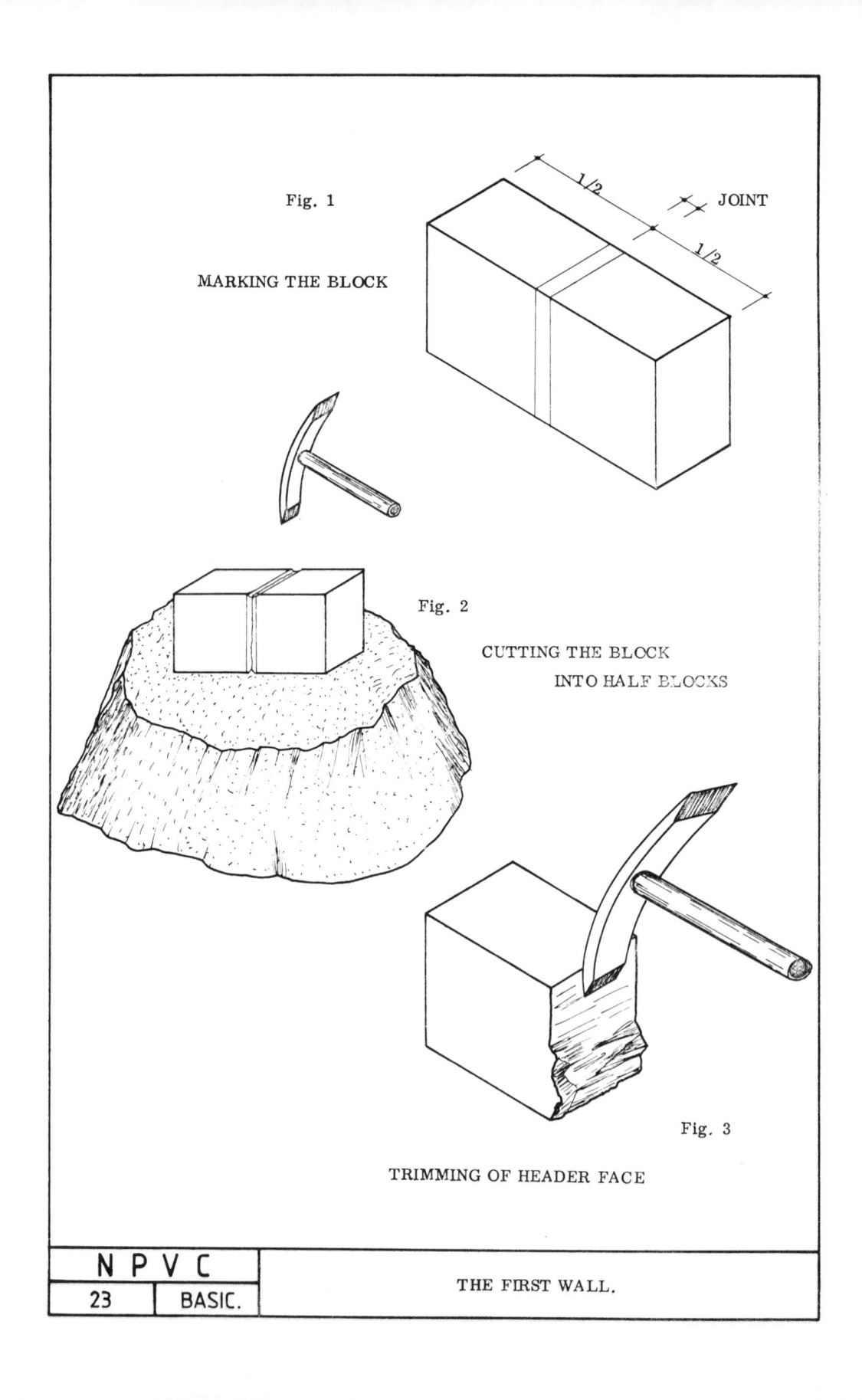

Fig. 1

MARKING THE BLOCK

JOINT
1/2
1/2

Fig. 2

CUTTING THE BLOCK INTO HALF BLOCKS

Fig. 3

TRIMMING OF HEADER FACE

N P V C	
23	BASIC.

THE FIRST WALL.

MARKING AND CUTTING BLOCKS

It is usually not possible to construct walls using only full blocks. In most cases 1/4, 1/2, and 3/4 blocks or intermediate sizes are also required. The previous description of the stopped end, for example, has shown the need for 1/2 blocks.

This does not mean that 1/2 blocks are obtained by simply cutting a full block into two identical halves. This is because the cross joint and its thickness must be considered.

Therefore, the 1/4, 1/2, and 3/4 blocks are actually that part of a full block minus half the thickness of the cross joint (Fig. 1).

Before cutting a block, mark the required size around all the faces. To prevent mistakes and to speed up the work, use a block gauge (see Rural Building Reference Book, Tools, page 12). Position the appropriate setting of the gauge against the block, then mark off the measurements on the block face using a pencil or a nail.

Set the marked block on a small heap of sand and then cut it with the block scutch by repeatedly and carefully knocking along the mark, making a groove in the surface. Direct the blows close to each other all around the four faces and continue until the block breaks apart along the groove (Fig. 2).

Trim the resultant rough header faces, if necessary, with the edge of your trowel blade (for landcrete blocks). When cutting sandcrete blocks, you should use the block scutch for trimming (Fig. 3).

- NOTE: Avoid cutting blocks on hard ground: they can easily break into irregular pieces and be wasted.

    Never cut blocks on a scaffold for the same reason, also because they might fall and injure someone below.

    It is always better to prepare in advance the number of blocks that you think will be needed that day.

    Do not use blocks which are cracked; these must be replaced by good blocks or the wall will be weakened.

NOTES:

THE FIRST WALL.

Fig. 1
ALIGN BLOCK WITH THE MARKS

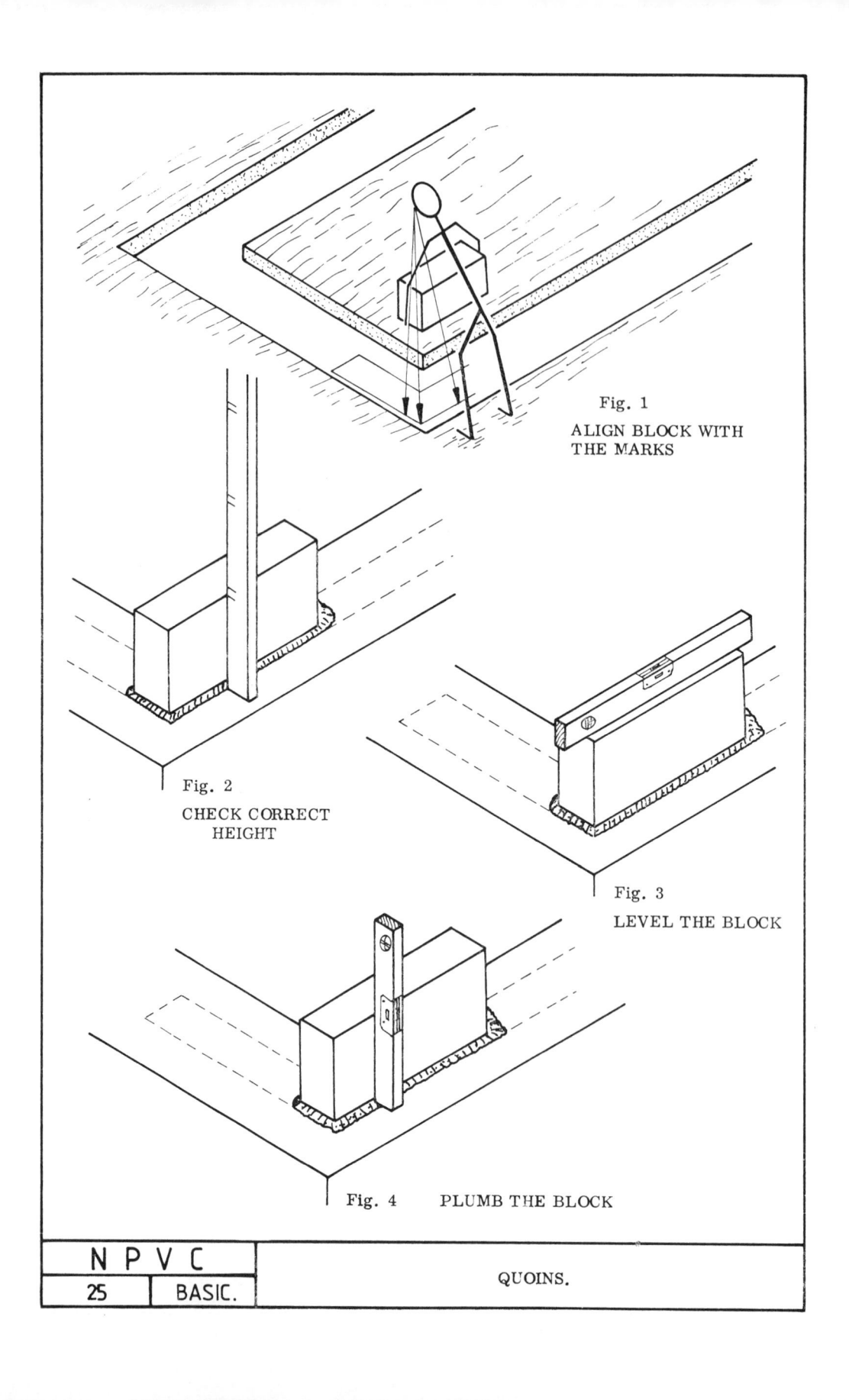

Fig. 2
CHECK CORRECT HEIGHT

Fig. 3
LEVEL THE BLOCK

Fig. 4   PLUMB THE BLOCK

N P V C	
25	BASIC.

QUOINS.

QUOINS

When external walls are constructed the corners or quoins are built first, to a height of several courses. Usually it is best to build six courses as this will reach to about 1,5 m high, the so-called scaffold height; and in most cases this will be half of the total height of the wall. The walling between the courses is completed later, course by course. The accuracy of the whole wall is determined by the corners, so great care must be taken to build them properly.

At the beginning of training, the positions of the quoins are determined by marking them out on the floor using the mason square.

A quoin is constructed in the following manner:

Blocks are sometimes not correctly shaped, so the first block or angle block must be chosen carefully so that all its faces are square to each other.

As you lay the angle block, stand close to the foundation with your head vertically over the block. You should be able to see that both outer faces of the block are aligned with the mark below (Fig. 1). After this the block has to be accurately levelled and plumbed.

Use the straight edge with gauge marks to ensure that the block is laid at the correct height. Hold the straight edge vertically against the block; the top edge of the block should correspond to the gauge mark (Fig. 2).

Now you have to make certain that the header face and the stretcher face are truly vertical. To do this hold the spirit level against one face about 5 cm from the corner, keeping it in this position while with your other hand you move the block until the bubble in the tube is centered. This operation must be repeated with the other face of the block (Figs. 3 & 4).

NOTES:

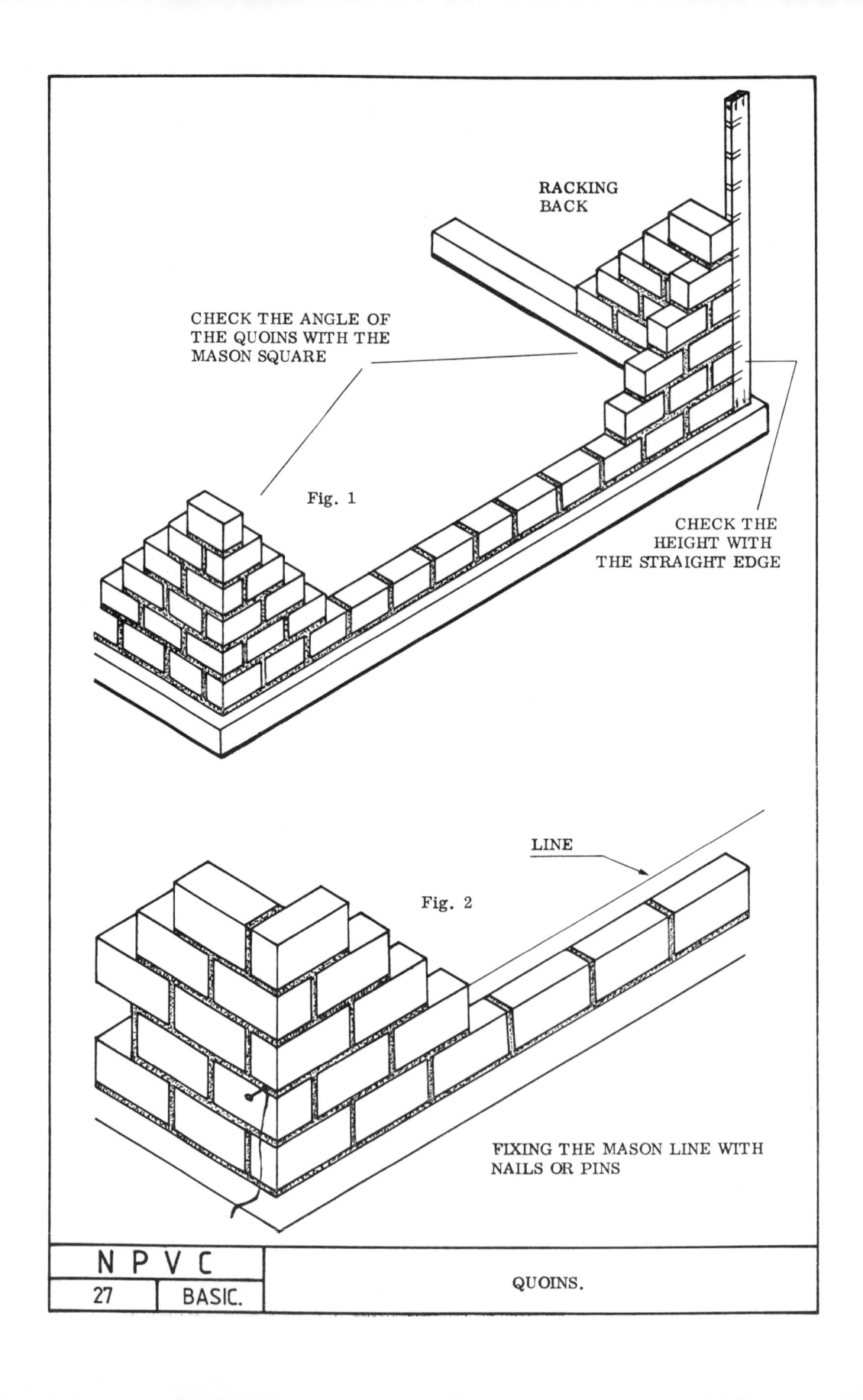

RACKING BACK

CHECK THE ANGLE OF THE QUOINS WITH THE MASON SQUARE

Fig. 1

CHECK THE HEIGHT WITH THE STRAIGHT EDGE

LINE

Fig. 2

FIXING THE MASON LINE WITH NAILS OR PINS

N P V C	
27	BASIC.

QUOINS.

Now lay several blocks in each direction according to the method described before. On the quoin stretcher side three more blocks should be laid; followed by four on the header side of the quoin. This will be a sufficient base for building up a height of six courses with either a toothed end or racking back (Fig. 1).

Use the mason square to make sure that the quoin has an angle of 90 degrees. Hold it against the quoin so that both of the blades fully touch the faces of the blocks. Repeat this operation after turning the square around, so that the direction of the blades is reversed.

SECOND COURSE

Following the 1-2-1 rule, the first block of the second course will not be the angle block; but the one covering the cross joint between the quoin stretcher and the adjoining stretcher. If the angle block were laid first it could get pushed out of position when the other blocks are laid. The correct method fixes the angle block in position by the cross joint between it and the first block. A further reason for this procedure is that the cross joint between the quoin headers and the adjoining stretchers has a different thickness than the other cross joints. This problem will be explained later when we come to bonding problems.

To complete the second course use the same method as described for the construction of the first course.

- NOTE: The arrangement in any course is repeated in the courses two above or two below it. Therefore only two alternating block arrangements are used.

WALLING BETWEEN QUOINS

When both corners of a wall have been built up to a height of six courses, it is necessary to fill in the blockwork between them. This is done with the aid of the mason line and either nails, pins, or line bobbins.

FIXING THE MASON LINE

If nails or pins are used, insert one of them in the bed joint at one corner so that the line will be level with the upper edge of the course (see Rural Building Reference Book, Tools, page 6). Fix the mason line to the nail or pin without using a knot, as shown in Fig. 2. This is so that later it can be easily removed.

Stretch the line taut to prevent any sagging, and push the second nail or pin into the corresponding bed joint in the opposite quoin. The line should now be level with the top of the course to be built; and about 2 mm or the thickness of a trowel

Fig. 1
LINE BOBBINS

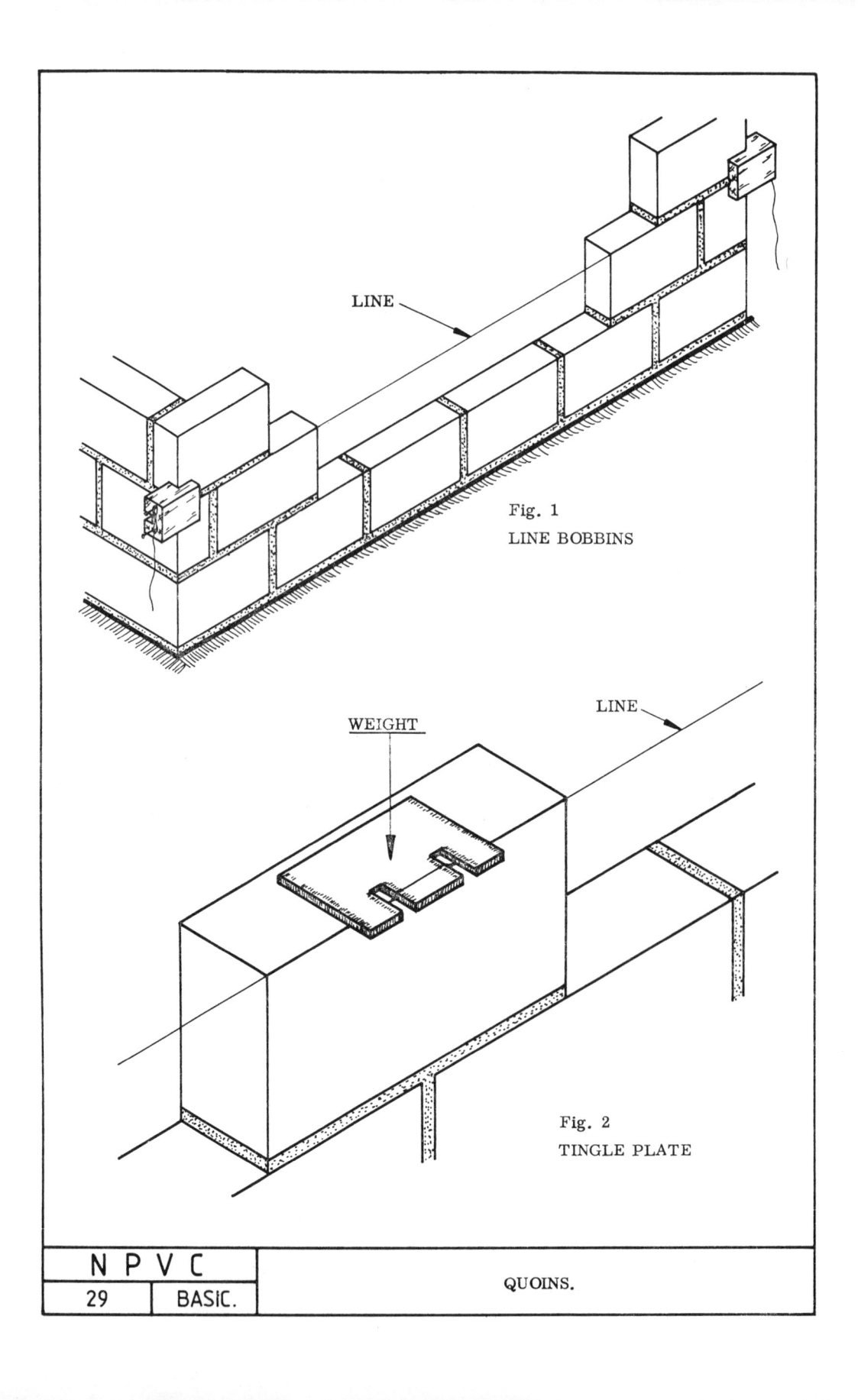

Fig. 2
TINGLE PLATE

N P V C	
29	BASIC.

QUOINS.

blade away from the blockwork. Put a wedge of paper between the wall and the line to keep the distance of 2 mm. The line should be horizontal.

If you use line bobbins instead of pins, take one with the line fastened around the screws and engaged in the saw cut; and position it with the notch against the corner of the quoin so that the line is level with the top edge of the course to be built. At the opposite quoin insert the line in the saw-cut of the second bobbin and set it at the correct height against the corner. Stretch the line taut and secure it by winding around the screws (Fig. 1).

The tight mason line holds the bobbins against the corners, keeping them in position. Once the line has been fixed in this manner no further adjustment is needed, unless the line starts to sag and needs tightening.

After one course is completed, simply slide the bobbins up the corners to the level of the next course.

USE OF THE TINGLE PLATE

If the line is stretched over a longer distance, it will tend to sag and will no longer provide a straight guide. In the case of a long wall where the distance between the bobbins exceeds 6 m, it becomes necessary to use one or more tingle plates (Fig. 2). This is done to keep the line from sagging. The tingle plate must be set on a so-called tingle block. Lay this block plumb, in position and at the correct height in the course to be built. This block keeps the tingle plate at the required height to support the mason line. Place the plate flat on the block, and weight it down with a half-block. The taut line is passed under the outer nibs and over the centre nib (Fig. 2).

COMPLETING THE COURSE

Now lay the blocks to complete the course. Take care that the outer top face edge of each block is level with the line.

At the same time be sure to keep the lower stretcher face edge in line with the edge of the course below. Avoid the tendency to lay the blocks too close to the line; the 2 mm distance must be maintained and checked from time to time by sliding the trowel blade between the line and block.

If the blocks are laid correctly, plumbing and levelling are not necessary. It is advisable however to check the face of the wall for a possible overhang, using the straight edge.

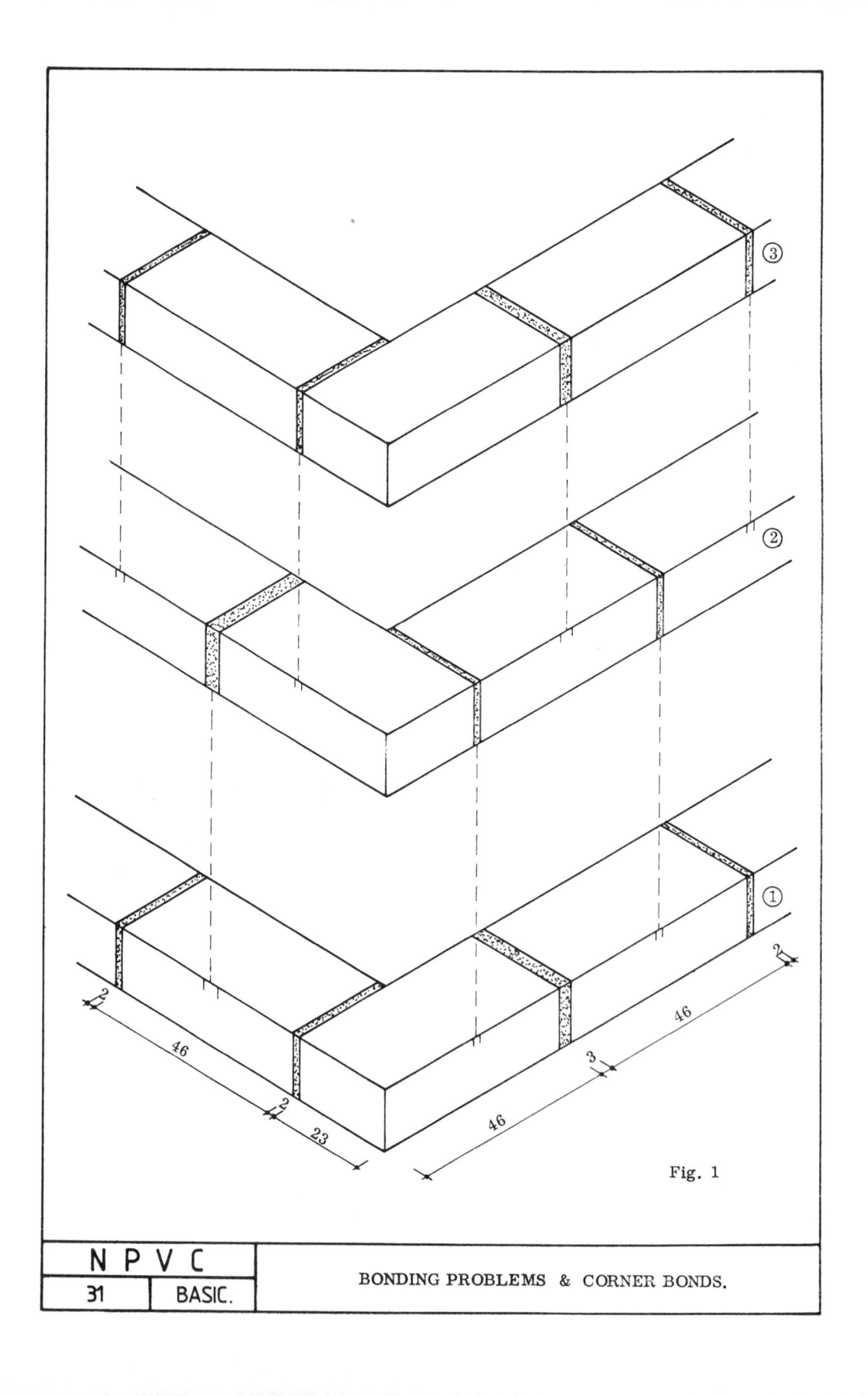

Fig. 1

N P V C	
31	BASIC.

BONDING PROBLEMS & CORNER BONDS.

After you have filled the cross joints, carefully rake out all the joints of the freshly laid courses to provide an additional grip for the plaster.

BONDING PROBLEMS

Problems arise when we use different types of blocks in the same wall. Usually in Rural Building we use sandcrete blocks for the lower courses of a wall to avoid problems with dampness, and landcrete blocks for the higher courses. Unfortunately the dimensions of sandcrete blocks and landcrete blocks do not match up well to each other. This means that it is not as easy to maintain a half-block bond as it is in a wall made up of only one kind of block.

For example, a sandcrete block has dimensions of 46 x 23 x 15 cm. A sandcrete half-block has the dimensions 22 x 23 x 15 cm. Even when the sandcrete blocks are cut in half they still will not match up exactly to the landcrete blocks (see the table below).

	Full block	Half block
Sandcrete	46 x 23 x 15	22 x 23 x 15
Landcrete	29 x 22 x 14	13,5 x 22 x 14

The dimensions of the 1/4 and 1/3 blocks also do not correspond, so in order to maintain the half-block bond between the sandcrete and landcrete courses we would have to make quite an adjustment.

Practically this means that although it would be better, we cannot maintain exactly a half-block bond between the sandcrete and landcrete. They must be considered as separate parts of the wall, properly bonded in themselves but not necessarily showing a half-block bond between them.

CORNER BONDS

CORNER BOND FOR FOOTINGS

The materials used for footings are sandcrete blocks which must be laid flatwise, giving a wall thickness of 23 cm. Since this measurement exceeds the length of a half-block by 1 cm, the first cross joints following the quoin stretchers must be 3 cm wide instead of the normal 2 cm. All the other joints are still 2 cm thick. By doing this the half-block bond is maintained (Fig. 1).

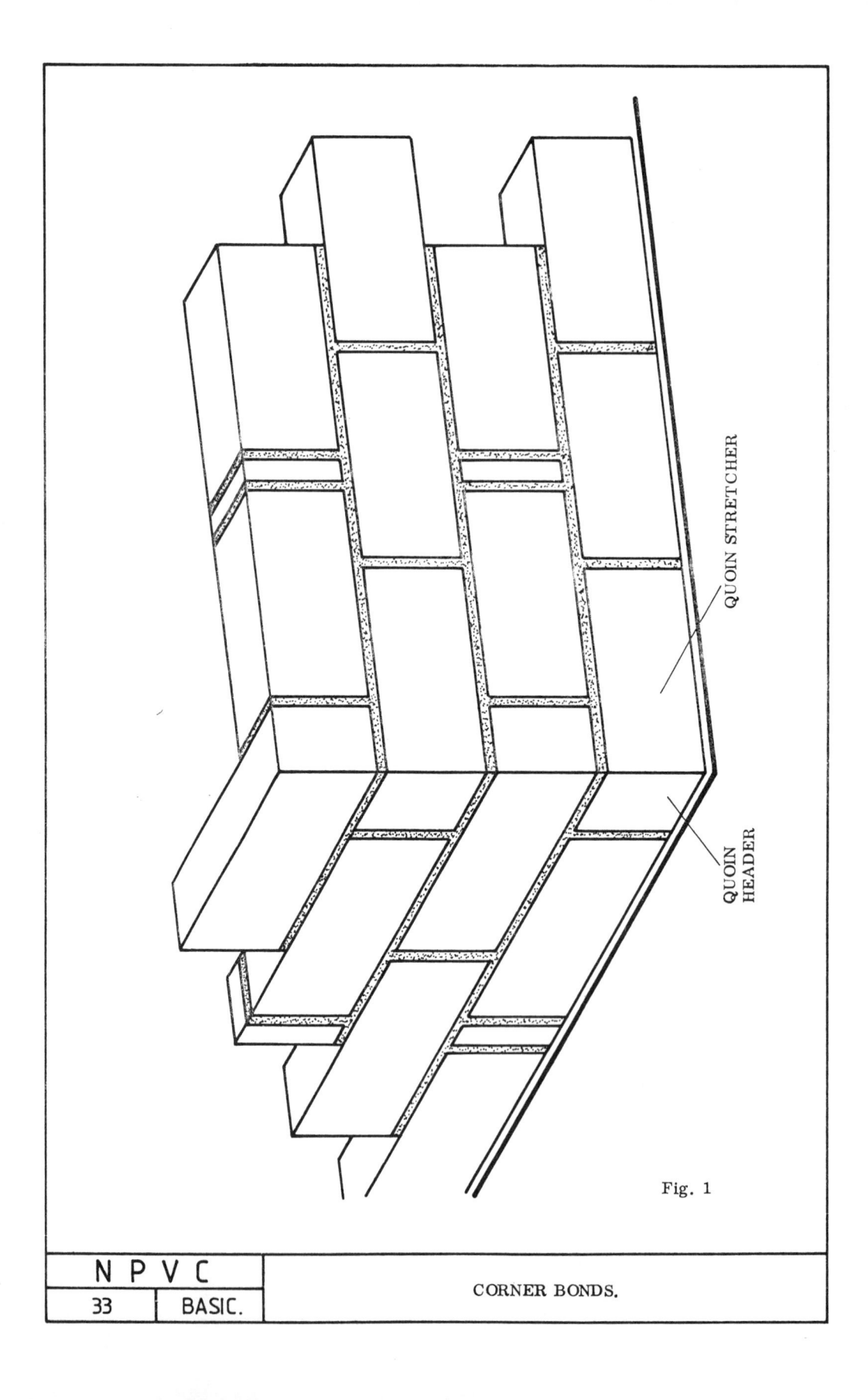

QUOIN STRETCHER

QUOIN HEADER

Fig. 1

N P V C	
33	BASIC.

CORNER BONDS.

There are of course two other ways to maintain the half-block bond:

- All the cross joints directly following the quoin header could be reduced to a thickness of only 1 cm. The effect would be the same but these joints would be difficult to fill properly, which could result in a weak quoin.
- All stretchers following the quoin headers could be cut to a length of 45 cm. This takes time and may damage the blocks.

In general, it is best not to use these last two methods.

CORNER BOND FOR A RISING SANDCRETE WALL

So-called wet rooms such as kitchens, showers and toilets must be built with sandcrete blocks. This is to avoid any damage to the walls caused by moisture.

If landcrete blocks are exposed to moisture for a long time the blocks will start to expand. This pushes the plaster off the walls and makes them weaker and weaker until finally they collapse under their own weight.

Rising walls in Rural Building are generally built by laying the blocks edgewise. When sandcrete blocks are used the wall has a thickness of 15 cm.

In order to avoid making too many cross joints within the quoin area, each quoin header as well as each quoin stretcher is followed by a full block.

To maintain the required half-block bond, a 7 cm lack of overlap has to be made up. This is done by inserting a 5 cm block between the first two stretchers that follow the quoin headers (5 cm plus 2 cm joint equals 7 cm). Almost every building project uses thin blocks for copings or rain gutters etc., and these specially made blocks can be simply cut in half and used to fill the gaps (Fig. 1).

NOTES:

CORNER BONDS.

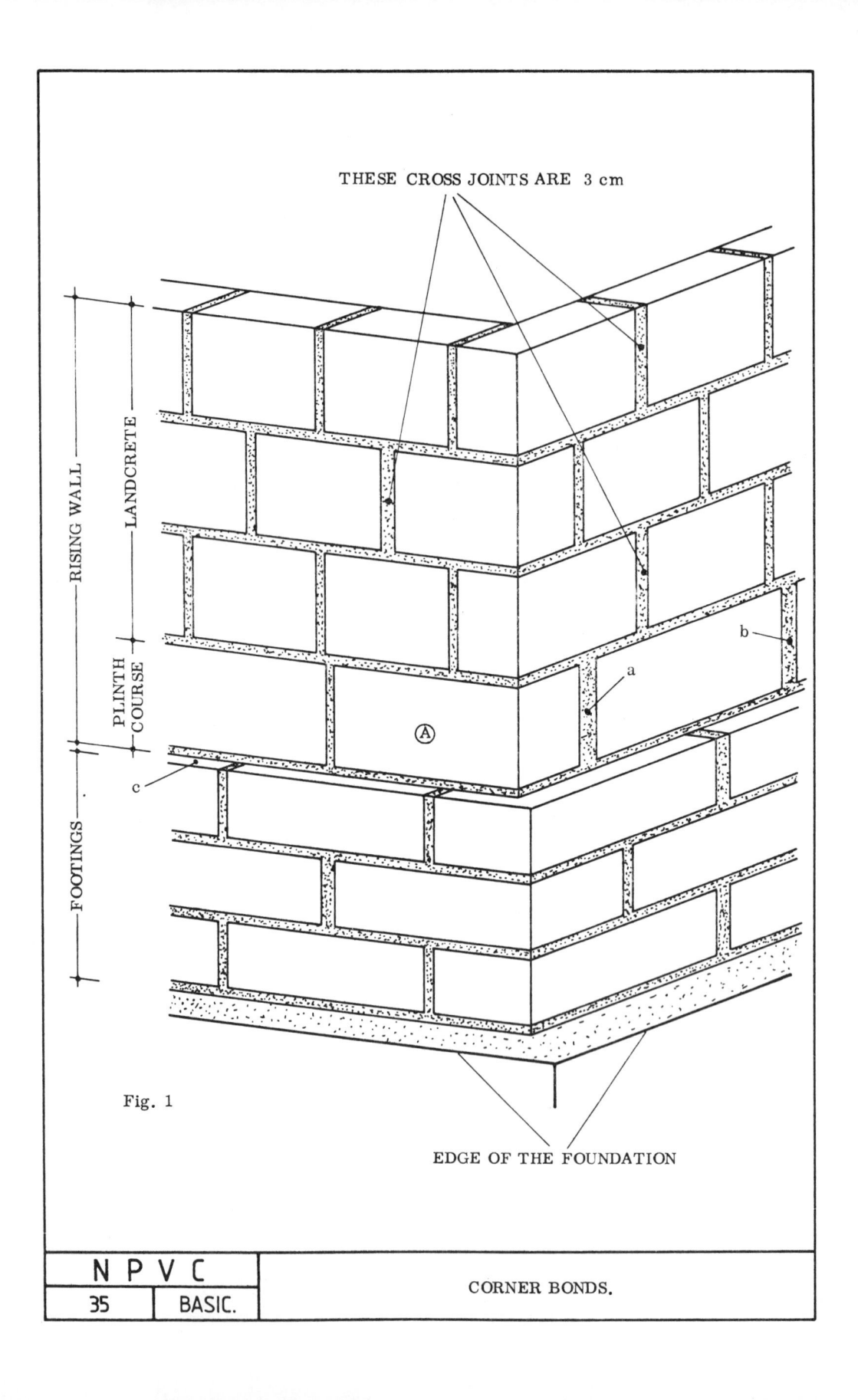

Fig. 1

THESE CROSS JOINTS ARE 3 cm

RISING WALL
LANDCRETE
PLINTH COURSE
FOOTINGS

a
b
c
A

EDGE OF THE FOUNDATION

N P V C	
35	BASIC.

CORNER BONDS.

CORNER BOND FOR RISING LANDCRETE WALL

As far as bonding is concerned, the footings, the plinth course and the rising landcrete wall are all regarded separately. They are properly bonded in themselves but don't necessarily show a half-block bond, especially between the plinth course and the landcrete wall.

As Fig. 1 illustrates, a half-block bond between footings and plinth course (both sandcrete) is possible despite the 4 cm setting back (c) and should be maintained.

- PLINTH COURSE: Since the rising wall is erected exactly in the middle of the footings, the plinth course has to be set back 4 cm from both faces: 23 cm (thickness of footings) minus 15 cm (thickness of plinth course) divided by 2 equals 4 cm (at both sides).

  The angle block (A) has to be shortened by 3 cm because of the setting back, thus we automatically make the correct half-block bond on the quoin stretcher side. The first two cross joints (a & b) following the quoin header must be widened to 4 cm each to overcome a lack of 4 cm in overlap. By doing this we distribute the lack of 4 cm over two joints equally.

- RISING LANDCRETE WALL: As already stated, the rising landcrete wall is regarded separately from the plinth course. This is because the dimensions of the landcrete and sandcrete blocks prevent the construction of a half-block bond between them.

  However, one important rule must be observed: No matter what part of the construction or what material it consists of, each quoin header must be overlapped by a quoin stretcher; each quoin stretcher is automatically followed by a quoin header (Fig. 1).

  To maintain the half-block bond within the landcrete wall, all the cross joints directly following the quoin stretchers must be widened to 3 cm. All other joints remain the same.

NOTES:

CORNER BONDS.

N P V C
BASIC. 36

CROSS SECTION OF
OUTSIDE WALL

F.V.

3cm

Ⓐ

Fig. 1   COMBINED SECTION OF OUTSIDE
         WALL AND VIEW OF INSIDE WALL

VIEW OF
INSIDE WALL

1/4
3/4
a = 3 cm thick
1/4
3/4
10 cm   a
Ⓒ   Ⓐ   Ⓑ
              1/4

Fig. 2

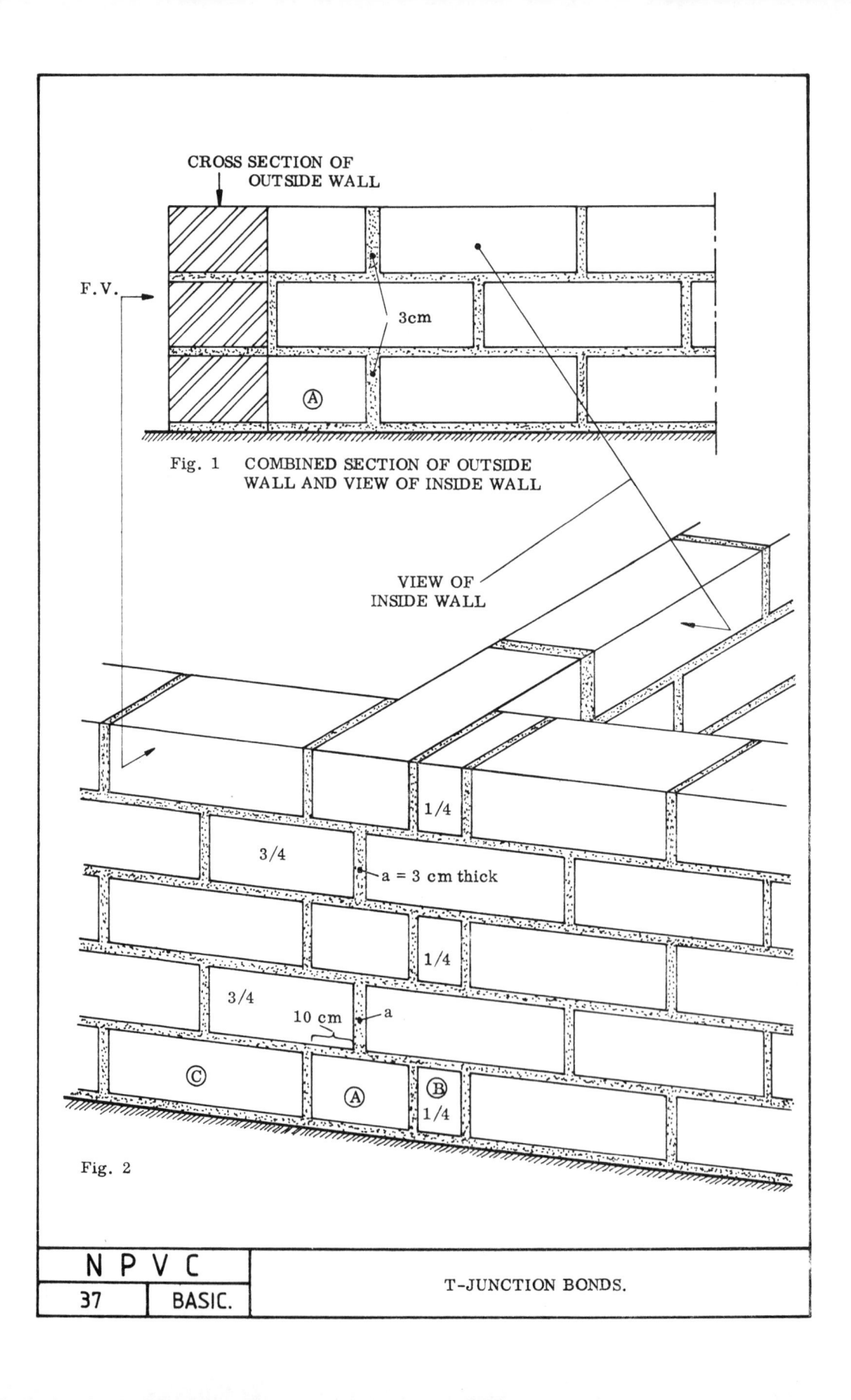

N P V C		
37	BASIC.	T-JUNCTION BONDS.

T-JUNCTION BONDS

The term T-junction is given to connections between walls which form a T shape, although it is not essential that the angles be right angles. This situation occurs most often where outside walls are met by inside walls.

T-JUNCTION BOND FOR FOOTINGS

Like quoins, the T-junctions are built first or at the same time as the quoins, and the walling between them is completed later.

The first block to be laid is the first block (A) of the inside wall: it will be seen as a header in the face of the outside wall (Figs. 1 & 2). This followed by a 1/4 block (B) at one side of it in the direction of the outside wall, and by a full block (C) on the other side (Fig. 2). This is followed by laying full blocks in all three directions.

The second course, and all the alternate courses, go through in the direction of the outside wall, that is, they do not share a block with the inside wall. Thus the inside wall is bonded to the outside wall only at alternating courses: at the 1st, 3rd, 5th etc. courses (Figs. 1 & 2).

The second course starts with a 3/4 block overlapping the header below by 10 cm but from the side opposite from the 1/4 block below. This is followed by full blocks in all three directions (Figs. 1 & 2).

The cross joints between the headers of the inside wall (a) are 3 cm thick and must be exactly in line with the centre of the headers (Fig. 1). The cross joints directly following the blocks which are bonded to the outside wall are also 3 cm thick (Fig. 2). All the other joints are 2 cm thick.

- NOTE: All courses with odd numbers (1st, 3rd, 5th etc.) share one block with the inside wall; these blocks are seen as headers in the face of the outside wall, each one next to a 1/4 block.

  All courses with an even number (2nd, 4th, 6th, etc.) go through in the direction of the outside wall and contain a 3/4 block at the opposite side of the 1/4 block below and above.

NOTES:

Fig. 1   2nd CONSTRUCTION

Fig. 2   3rd CONSTRUCTION

Fig. 3   4th CONSTRUCTION

N P V C	
39	BASIC.

T-JUNCTION BONDS.

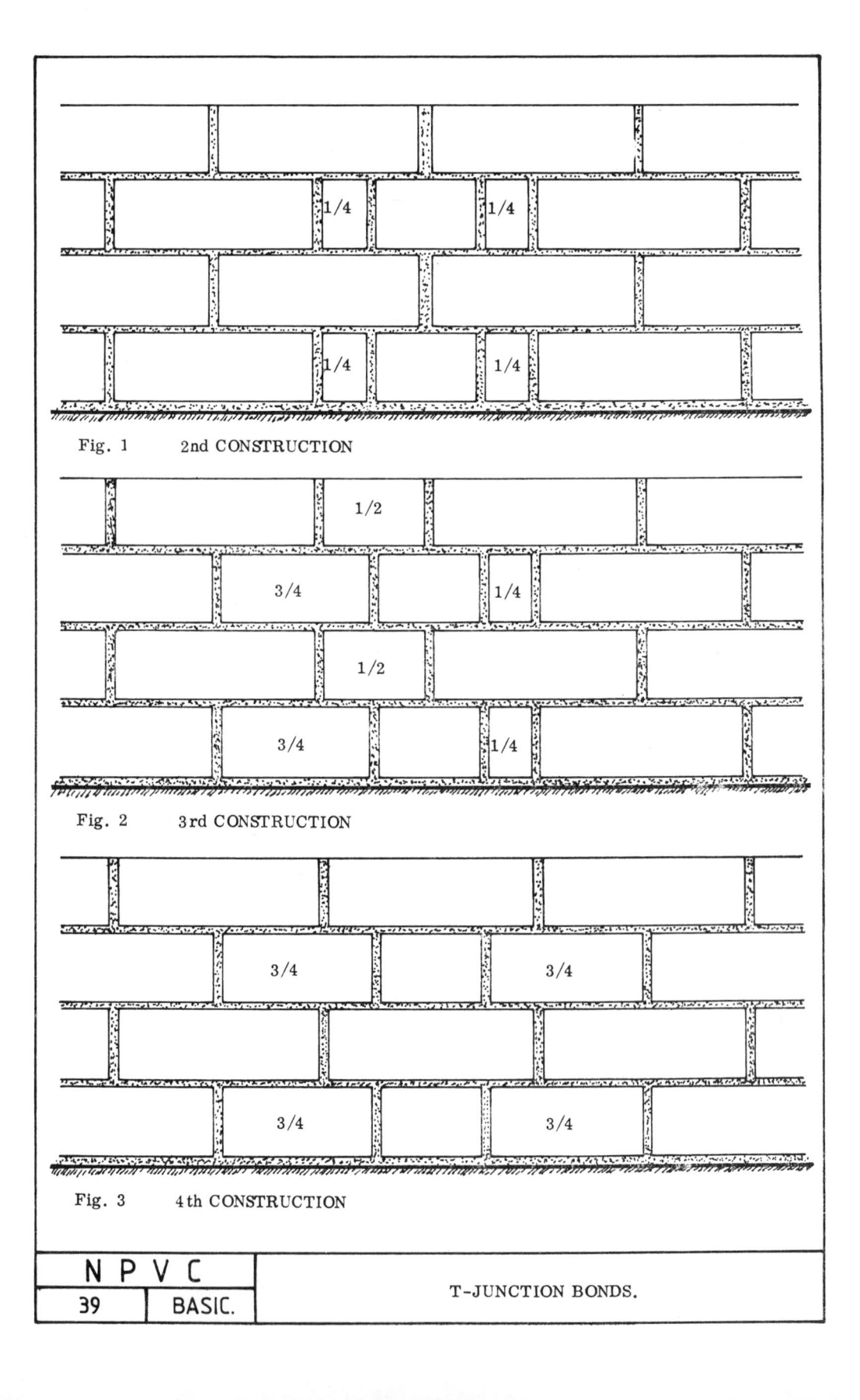

The illustrations on the opposite page show three other possible constructions for T-junctions in footings, all of which maintain the half-block bond (Figs. 1, 2, & 3).

Which method is used depends on factors such as the distance between the junction and the next quoin, or the next junction.

One way to find out the best choice is to lay out the first course of blocks without mortar, and try different arrangements. In this way the builder can decide what will be the best final arrangement.

If two or more constructions are possible the Rural Builder should chose the most efficient method. This is the method that wastes the least time and materials.

Specifically, the more blocks that have to be cut the more time will be needed; and if for example only 3/4 blocks are needed, the remaining 1/4 blocks are wasted unless there is a need for them somewhere else.

By comparing the four possible constructions (Figs. 1 & 2 on page 37, and Figs. 1, 2 & 3 at left) we see that the first three can be carried out without any waste of blocks while in the last case (Fig. 3) all the 1/4 blocks are left-over.

As far as efficient work is concerned, we see that for the first method (page 37) only one cut needs to be made for two courses as both parts of the block are used. The alternatives shown in Figs. 1, 2 and 3 need two cuts for two courses.

By considering carefully the advantages and disadvantages of the four, we see that the first construction is the best, the 2nd and 3rd constructions are less good and the 4th construction is the worst. Therefore the first type should be used whenever possible.

NOTES:

T-JUNCTION BONDS.

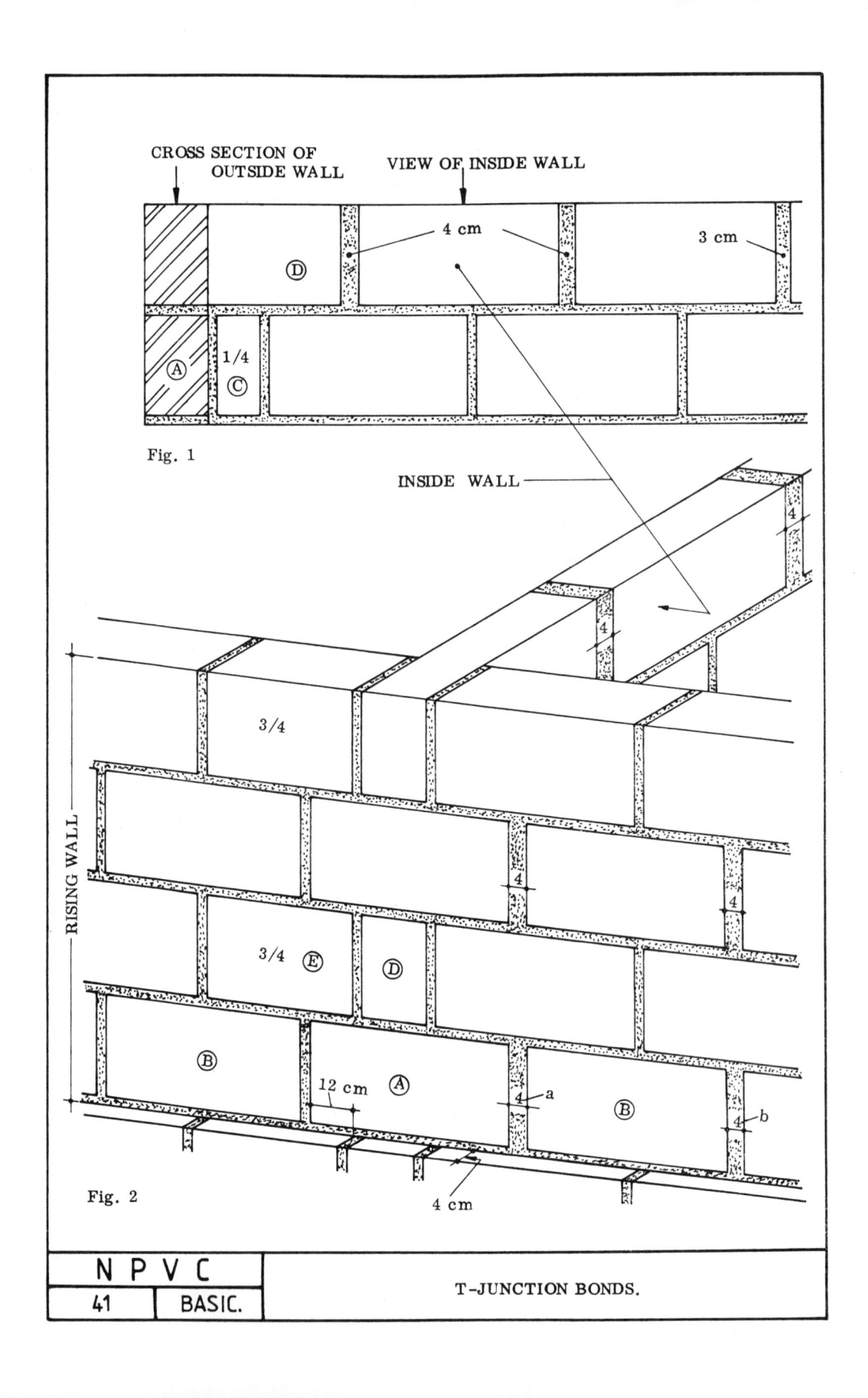

Fig. 1

CROSS SECTION OF OUTSIDE WALL
VIEW OF INSIDE WALL
4 cm
3 cm
Ⓓ
Ⓐ
1/4 Ⓒ

Fig. 2

INSIDE WALL
RISING WALL
3/4
4
3/4 Ⓔ Ⓓ
4
4
Ⓑ
12 cm Ⓐ 4 a Ⓑ 4 b
4 cm

N P V C
41 BASIC.
T-JUNCTION BONDS.

T - JUNCTION BOND FOR RISING SANDCRETE WALLS

Since the footings are normally 3 courses high, the third course of the footings contains the block which bonds the inside wall. Thus the first course of the rising sandcrete wall must go through in the direction of the outside wall to cover the header of the footings (see previous pages).

The first block of the rising wall is placed so that it extends past the header below by 12 cm on one side (Fig. 2, block A). Don't forget to set this block back from the face of the footings by 4 cm. This block is followed by full blocks on either side (Fig. 2, blocks B). The inside wall starts with a 1/4 block (Fig. 1, block C), this too is followed by full blocks.

The second course of the T-junction begins with the full block of the inside wall that is bonded into the outside wall (Figs. 1 & 2, block D). This is followed by a 3/4 block (Fig. 2, block E) on the same side of the header where the stretcher (block A) below projects by 12 cm. Continue with full blocks in all three directions.

The odd-numbered courses of the outside wall contain widened cross joints to catch up with the half-block bond. The first two cross joints opposite the 12 cm projection (Fig. 2, a & b) are 4 cm thick, followed by one more joint which is 3 cm thick (not shown). In this way the half-block bond is maintained.

The same must be done with the first three cross joints following the bonded block of the inside wall (Fig. 1).

NOTES:

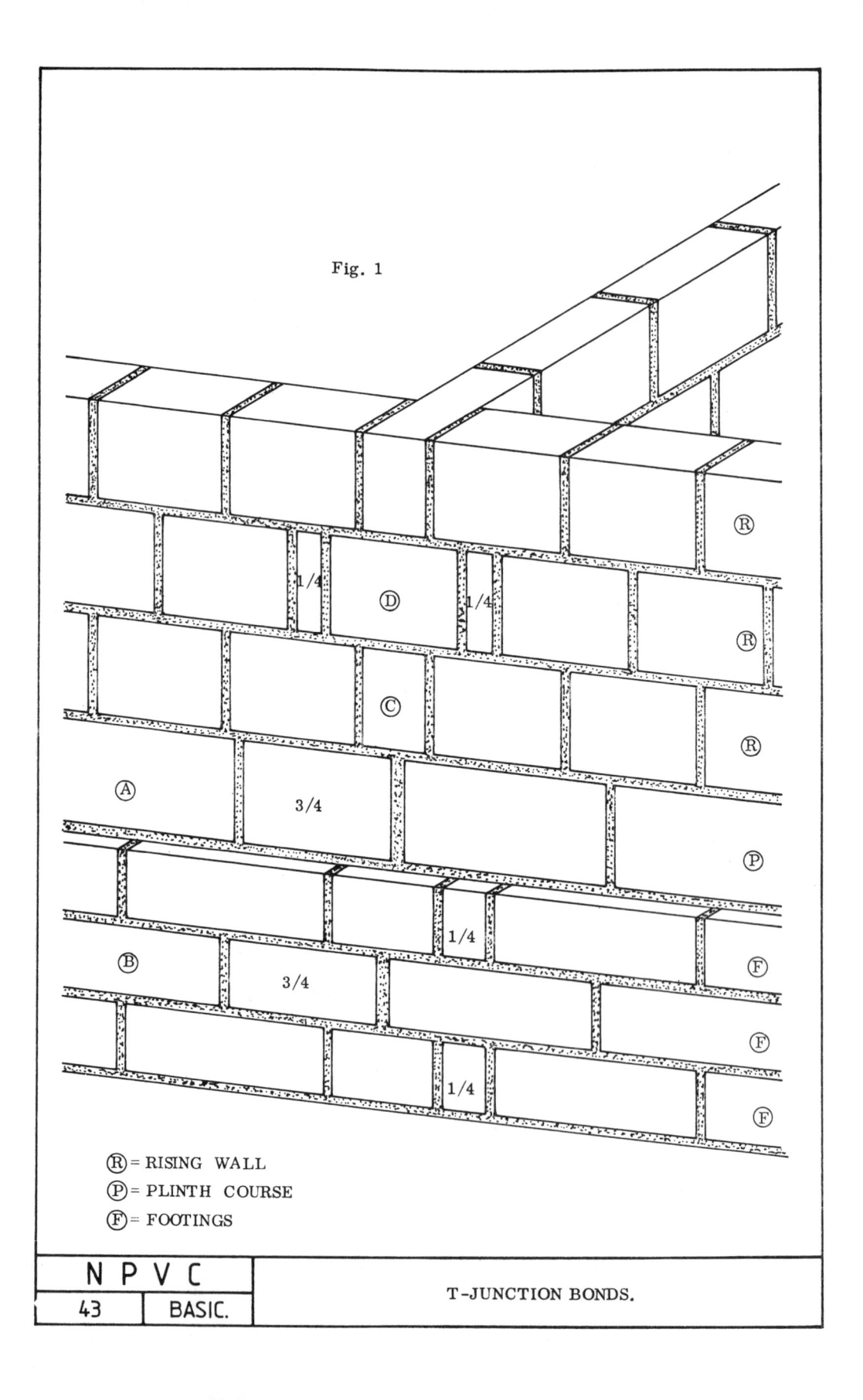

Fig. 1

Ⓡ = RISING WALL
Ⓟ = PLINTH COURSE
Ⓕ = FOOTINGS

N P V C	
43	BASIC.

T-JUNCTION BONDS.

T-JUNCTION BOND FOR RISING LANDCRETE WALLS

- PLINTH COURSE: Except that the blocks are laid edgewise rather than flat, the plinth course is simply a repetition of the footing course two courses below (Fig. 1, blocks A & B).

    The only difference is that the first two cross joints in the inside wall are widened to 4 cm each, in order to maintain the half-block bond in relation to the footings.

    Do not forget to set the plinth course back from the face of the footings by 4 cm.

- RISING LANDCRETE WALL: Since the landcrete wall is 1 cm thinner than the plinth course, it is essential to continue the "good" face of the plinth by setting the landcrete blocks flush with the outside face of the plinth. The Rural Builder should choose one face of the inside wall to be the "good" face; where the surface of the plinth course is flush to that of the rising wall. This is the face from which the plumbing and levelling are done; the Rural Builder should alway work from this face. On outside walls the "good" face is normally the outside face.

    The first block of the landcrete wall is the one which is shared by the inside and outside walls and covers two cross joints (Fig. 1, block C). It is followed in all three directions by full blocks.

    The second course of the rising landcrete wall begins with a full block centred exactly over the header below, with 1/4 blocks on either side (block D). All of the other courses in the rising wall are repetitions of these two courses.

NOTES:

Fig. 1

Ⓐ

2,5 cm

Fig. 2

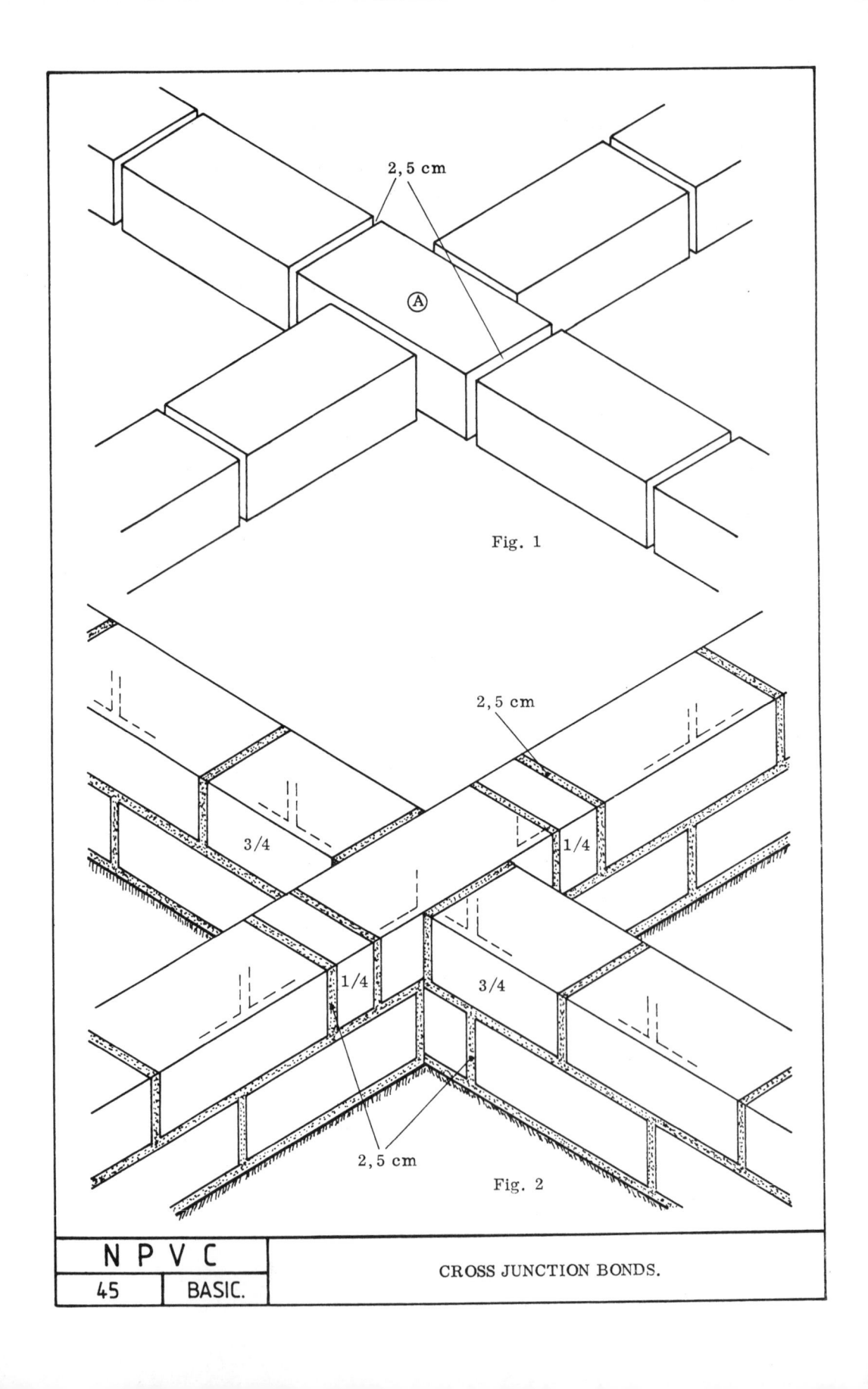

2,5 cm

3/4 1/4

1/4 3/4

2,5 cm

N P V C	
45	BASIC.

CROSS JUNCTION BONDS.

CROSS JUNCTION BONDS

A cross junction, also called an intersection, consists of two continuous walls which intersect, or cross each other.

The following are only a few examples out of many methods for bonding at a cross junction. In actual practice the dimensions of the building will not always permit the use of these particular bonds. In such cases some adjustments must be made.

The essential requirements for a cross junction always remain the same:

- avoid making continuous cross joints; and
- try to use a minimum number of cut blocks.

CROSS JUNCTION BOND FOR FOOTINGS

The first course consists entirely of full blocks (Fig. 1). The through-going block (A) projects equally from both sides of the crossing wall. This block has cross joints of 2,5 cm on each end, but all the other joints are still 2 cm thick (Fig. 1).

The second course starts with a full block centred over the through-going block below. This block is followed by 1/4 blocks on either end. The 1/4 blocks each have one cross joint which is 2,5 cm thick. All other joints are still 2 cm thick.

The second course of the crossing wall continues with 3/4 blocks on both sides. This is followed by full blocks in all four directions.

NOTES:

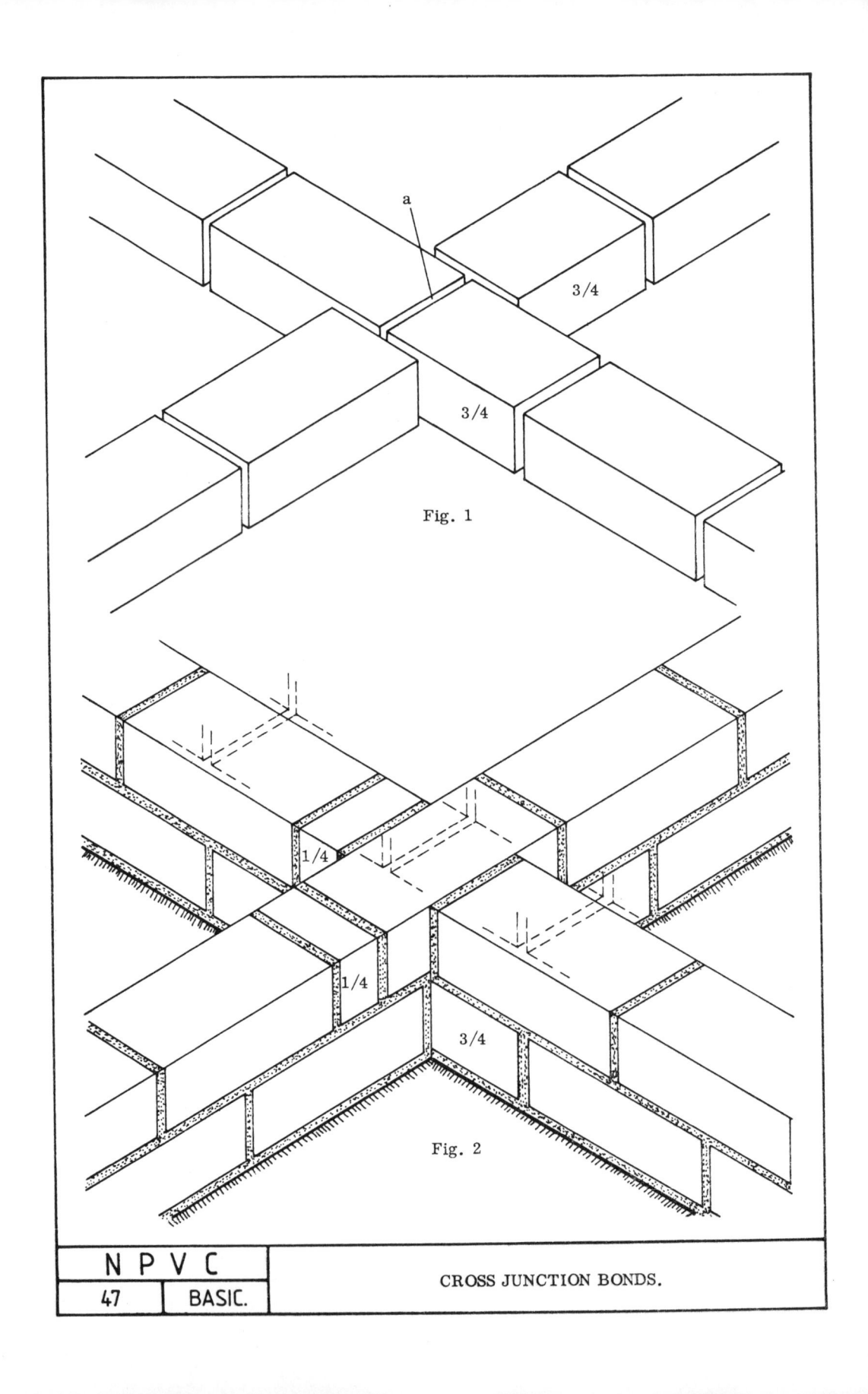

Fig. 1

Fig. 2

N P V C		CROSS JUNCTION BONDS.
47	BASIC.	

The second possible construction is where the crossing walls enclose a cross joint (Fig. 1, a). In this case two 3/4 blocks are used in the first course while the remaining 1/4 blocks are kept aside for the second course (Fig. 1).

The two 3/4 blocks of the first course are arranged so that they are at right angles to each other. One of the 3/4 blocks always starts from the middle of the crossing wall while the other is placed on either the right or left side of it. This is followed in all four directions by full blocks (Fig. 1).

The second course begins with a full block crossing the through-going wall below and overlapping equally at both sides. This block covers three cross joints at once instead of two as in the last method. The first block is followed by the 1/4 blocks placed on opposite sides from the 3/4 blocks below. The second course is continued with full blocks in all four directions (Fig. 2).

Each 1/4 block must have one cross joint of 2,5 cm; all other joints remain the same, 2 cm thick.

Try to figure out more possible ways to do a cross junction bond, and discuss the results with your fellow trainees, your instructors, and your foreman on the building site.

NOTES:

Fig. 1

Fig. 2

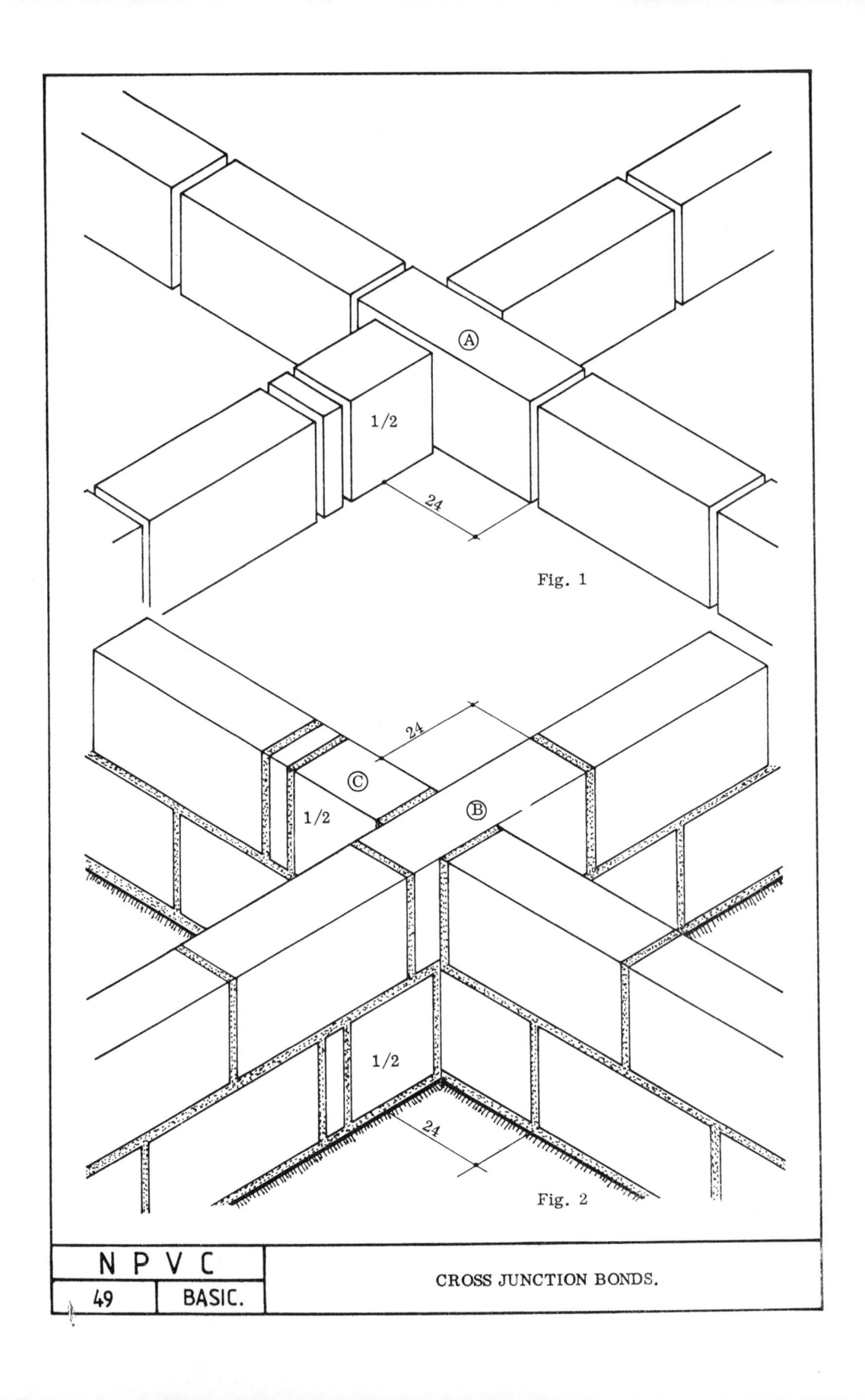

N P V C		CROSS JUNCTION BONDS.
49	BASIC.	

CROSS JUNCTION BOND FOR RISING SANDCRETE WALLS

The smaller wall thickness in relation to the block length makes it necessary that all cross junction bonds for rising sandcrete walls contain a block that is 5 cm thick (compare with corner bonds for rising sandcrete walls, page 34).

The opposite illustrations show the most economical method for constructing a cross junction. Apart from the one 1/2 block and one 5 cm block, both the first and second courses contain only full blocks.

Fig. 1 shows the first course. The through wall consists of only full blocks, while the crossed wall starts on one side with a 1/2 block combined with a 5 cm block. This is continued in all four directions with full blocks. The through-going block (A) must extend past the crossed wall by 24 cm, which is the length of a 1/2 block plus the joint.

In the second course, the wall which was crossed in the course below now goes through. The through block (B) again extends past the crossed wall by 24 cm, but on the side opposite from the 1/2 block in the course below. Block B is followed on its left-hand side by a 1/2 block (C) combined with a 5 cm block. This is continued in all four directions with full blocks (Fig. 2).

In the first course the crossing block and the 1/2 block form a corner. This is also true in the second course but the corner is diagonally opposite from the one in the first course below.

- REMEMBER: All through courses in either wall consist of full blocks only.
  All crossed courses in either wall contain one 1/2 block combined with a 5 cm block.

NOTES:

CROSS JUNCTION BONDS.

N P V C
BASIC. 50

Fig. 1

Fig. 2

N P V C		
51	BASIC.	CROSS JUNCTION BONDS.

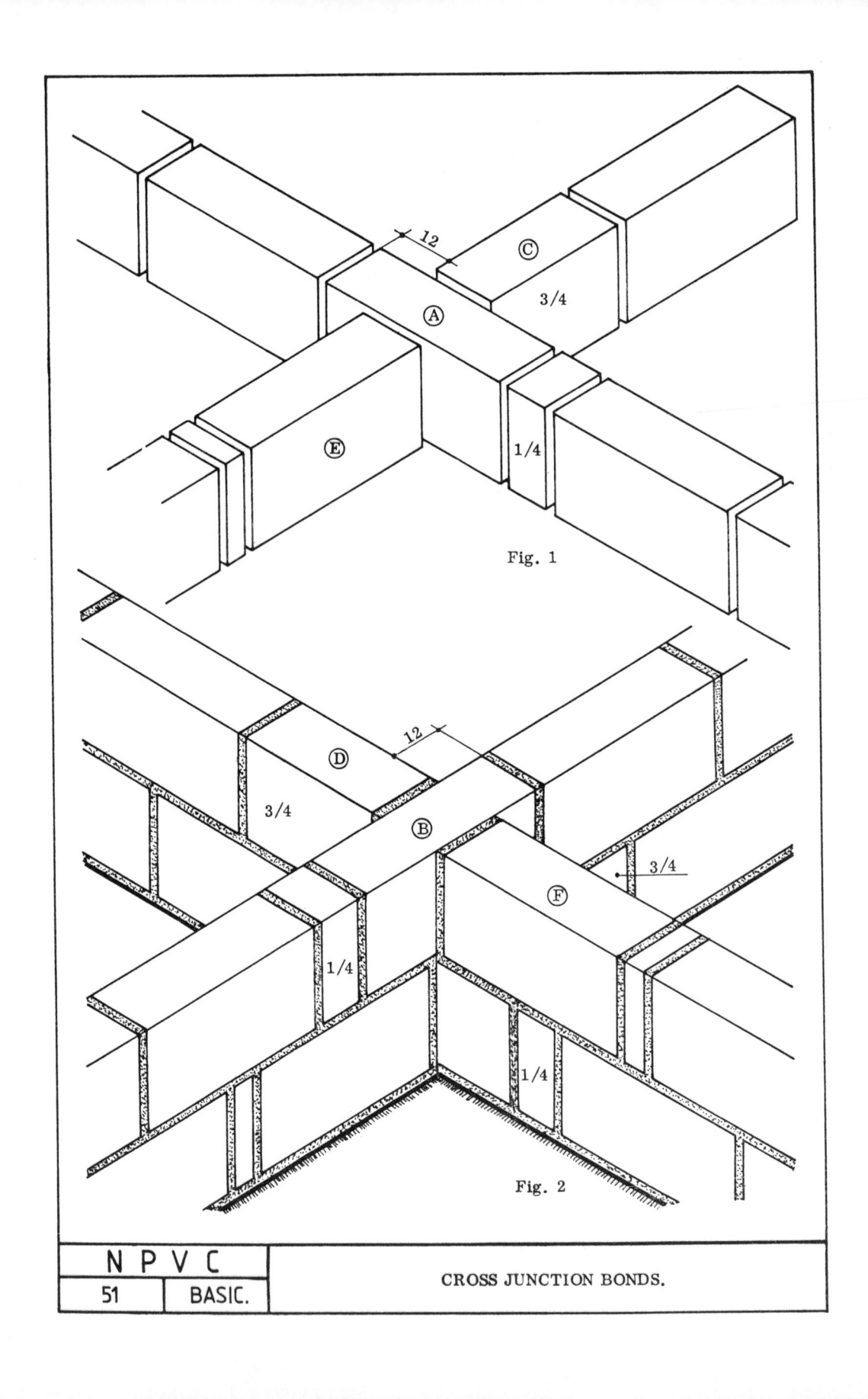

The second method of making a cross junction in a sandcrete wall is different from the first method because 3/4 and 1/4 blocks are used instead of 1/2 blocks.

Because of this the rules for this bonding method are also different:

- All through blocks (Figs. 1 & 2, blocks A & B) project from the crossed wall by 12 cm at one end (12 cm is 1/4 block plus the joint). These blocks are followed by 1/4 blocks at the other end (Figs. 1 & 2).

- All crossed walls contain a 3/4 block (Figs. 1 & 2, blocks C & D) set directly against the through-going block. This 3/4 block is on the opposite side from the 1/4 blocks in the course below (Fig. 1). The crossed wall continues with a full block (Figs. 1 & 2, blocks E & F) and a 5 cm block, on the other side from the 3/4 block.

- NOTE: The above method involves a lot of block-cutting, which makes it less efficient than the other method.

Try to figure out more possibilities and discuss them with your fellow trainees, your instructors and your foreman on the building site.

NOTES:

Fig. 1

Fig. 2

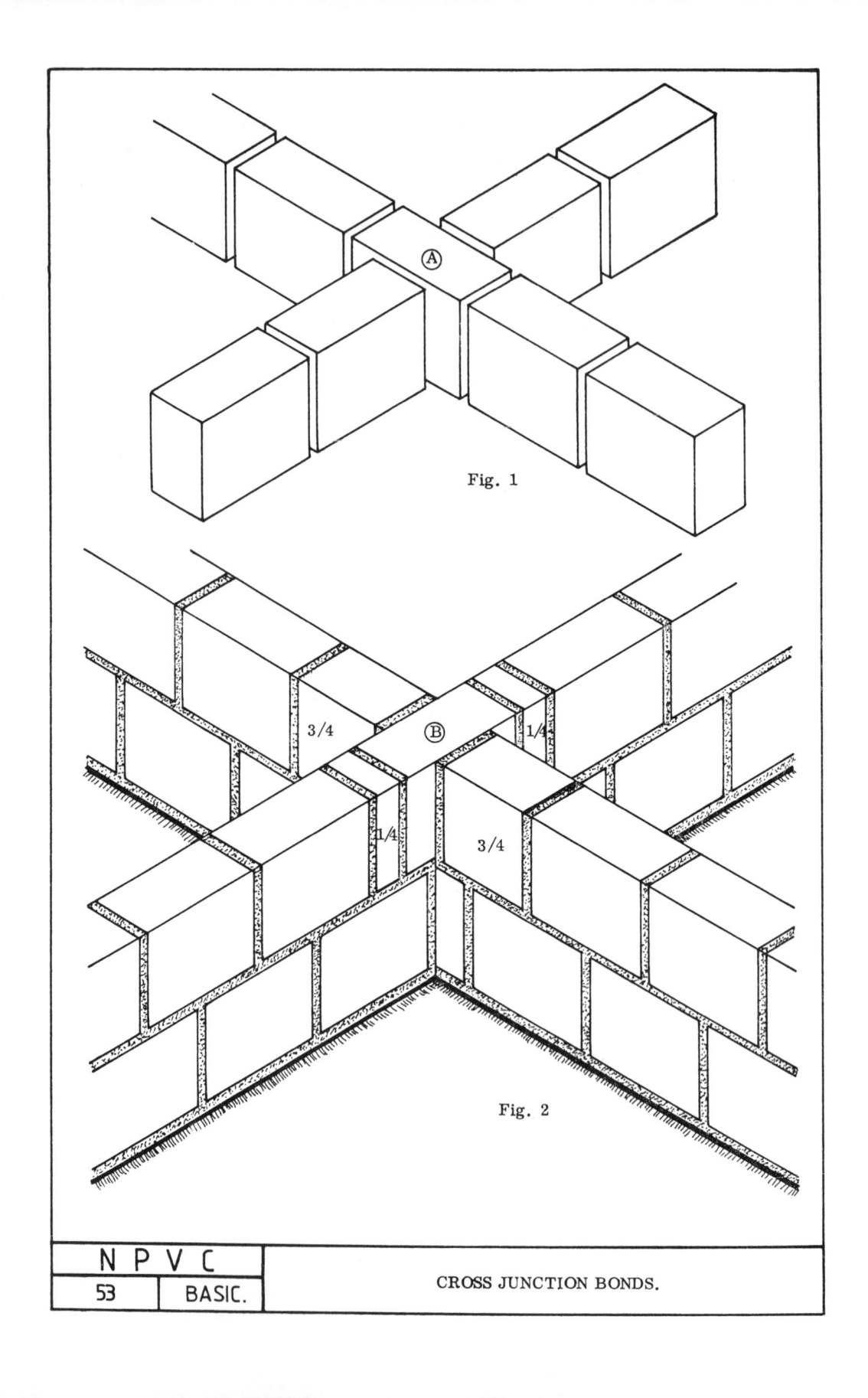

N P V C	
53	BASIC.

CROSS JUNCTION BONDS.

CROSS JUNCTION BOND FOR RISING LANDCRETE WALLS

The arrangement of blocks to construct a cross junction with landcrete blocks is almost the same as for the footings (see previous pages).

Fig. 1 shows the bond for all courses with an odd number (1st, 3rd, 5th, etc.). These consist of full blocks only. The through block (A) is set exactly in the middle of the crossed wall.

The arrangement for courses with an even number (2nd, 4th, 6th, etc.) is shown in Fig. 2. These courses also start with a full block (B) set exactly across the middle of the through-going block below. This block is followed by 1/4 blocks on its ends and 3/4 blocks on its stretcher sides. This is continued in all four directions with full blocks.

NOTES:

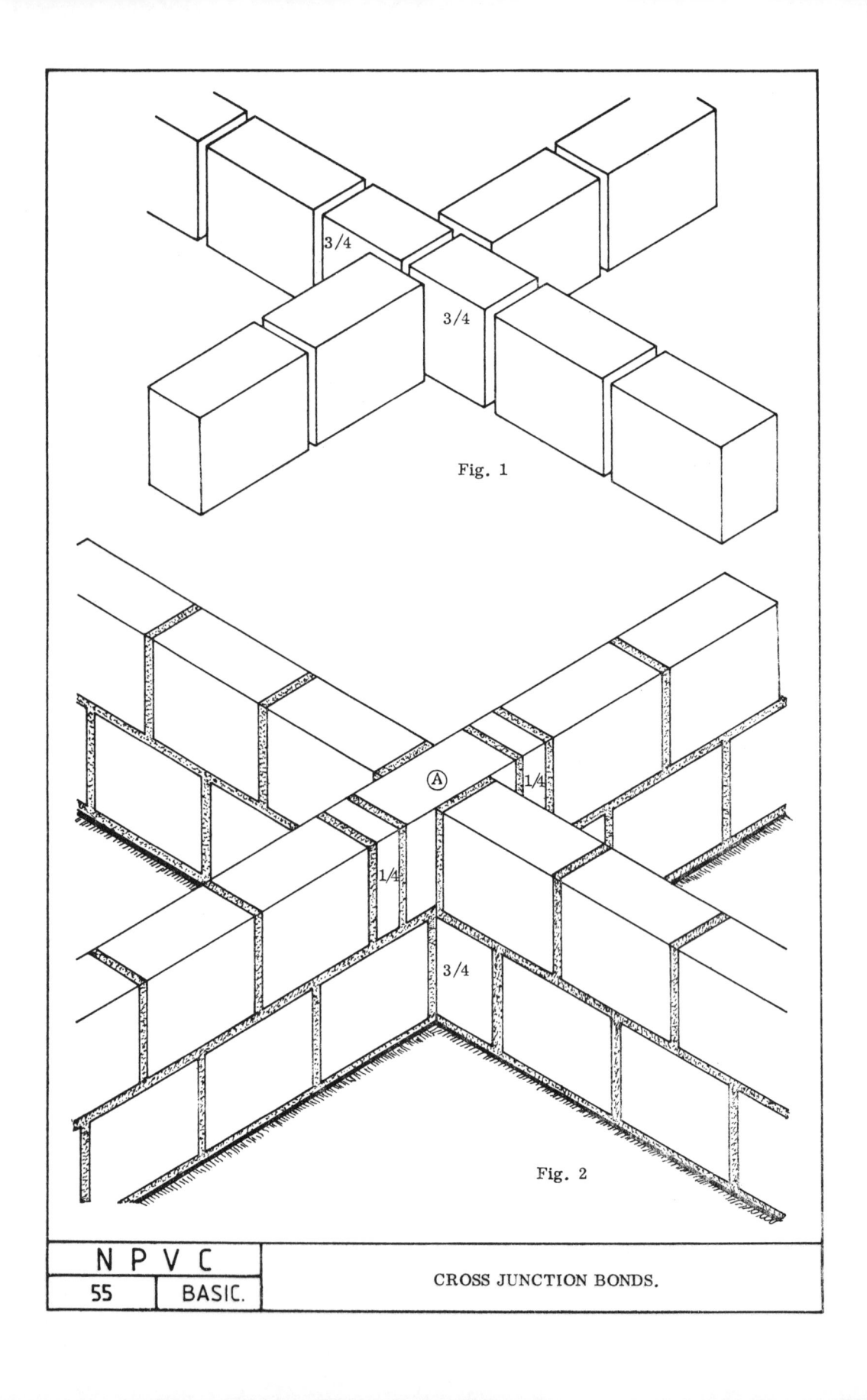

Fig. 1

Fig. 2

NPVC
55 BASIC.

CROSS JUNCTION BONDS.

Another method of bonding uses the cut blocks in alternate courses, unlike the last method which used all cut blocks in the same course. There is no waste in either method.

The first crossing (Fig. 1) is formed by two 3/4 blocks which project equally from the crossed wall, meaning that the joint between them is exactly in the centre of the crossed wall. The course is continued in all four directions with full blocks.

The second course begins with a through-going block (Fig. 2, block A) set exactly in the middle of the crossed wall. In this way, three cross joints are covered by the block, instead of two as in the last method (Fig. 2).

The through-going block is followed by 1/4 blocks on its ends but then the course continues in all directions with full blocks.

Try to develop more possibilities but do not forget the requirements mentioned on page 46.

NOTES:

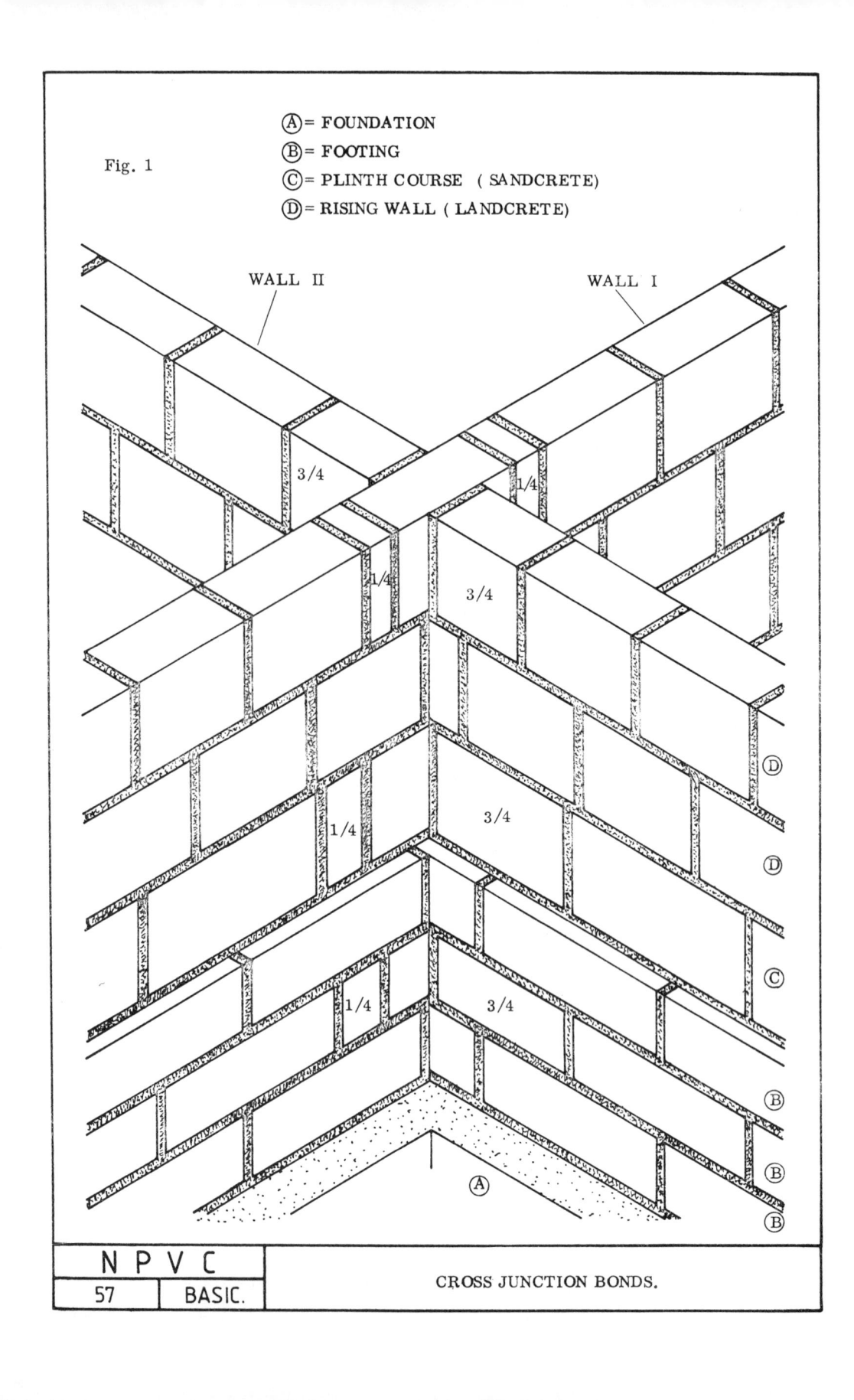

Fig. 1

Ⓐ = FOUNDATION
Ⓑ = FOOTING
Ⓒ = PLINTH COURSE (SANDCRETE)
Ⓓ = RISING WALL (LANDCRETE)

WALL II   WALL I

3/4   1/4
1/4   3/4
1/4   3/4
1/4   3/4

Ⓓ
Ⓓ
Ⓒ
Ⓑ
Ⓑ
Ⓐ
Ⓑ

NPVC
57 BASIC.

CROSS JUNCTION BONDS.

FOOTINGS - PLINTH COURSE - RISING LANDCRETE WALL

The illustration opposite shows what a cross junction will look like during the construction (Fig. 1).

Again, it is possible to maintain the half-block bond between the footings and the plinth course in wall I, as described on the previous pages. The bonding of the plinth course is a repetition of the bonding of the second footing course.

However, it can be seen that the half-block bond between the footings and the plinth course is not perfectly maintained in wall II. This is because of the 4 cm setting back on both sides as well as the reduced wall thickness.

The bond of the rising landcrete walls is the same as described on the previous pages. Only two courses are shown because the rest of the courses are repetitions of these two.

If you compare the bond of the rising landcrete wall with the plinth course, you will see that the half-block bond is not maintained between them. The only way it could be maintained is by cutting more blocks and changing the thicknesses of some of the joints. This would weaken the entire wall as well as wasting materials, so the Rural Builder should consider this bonding all right as it is shown here.

- REMEMBER: The rising landcrete wall is 1 cm thinner than the plinth course. Do not forget to maintain the "good" faces of the walls by laying the landcrete blocks flush with the plumbed faces of the plinth courses below.

NOTES:

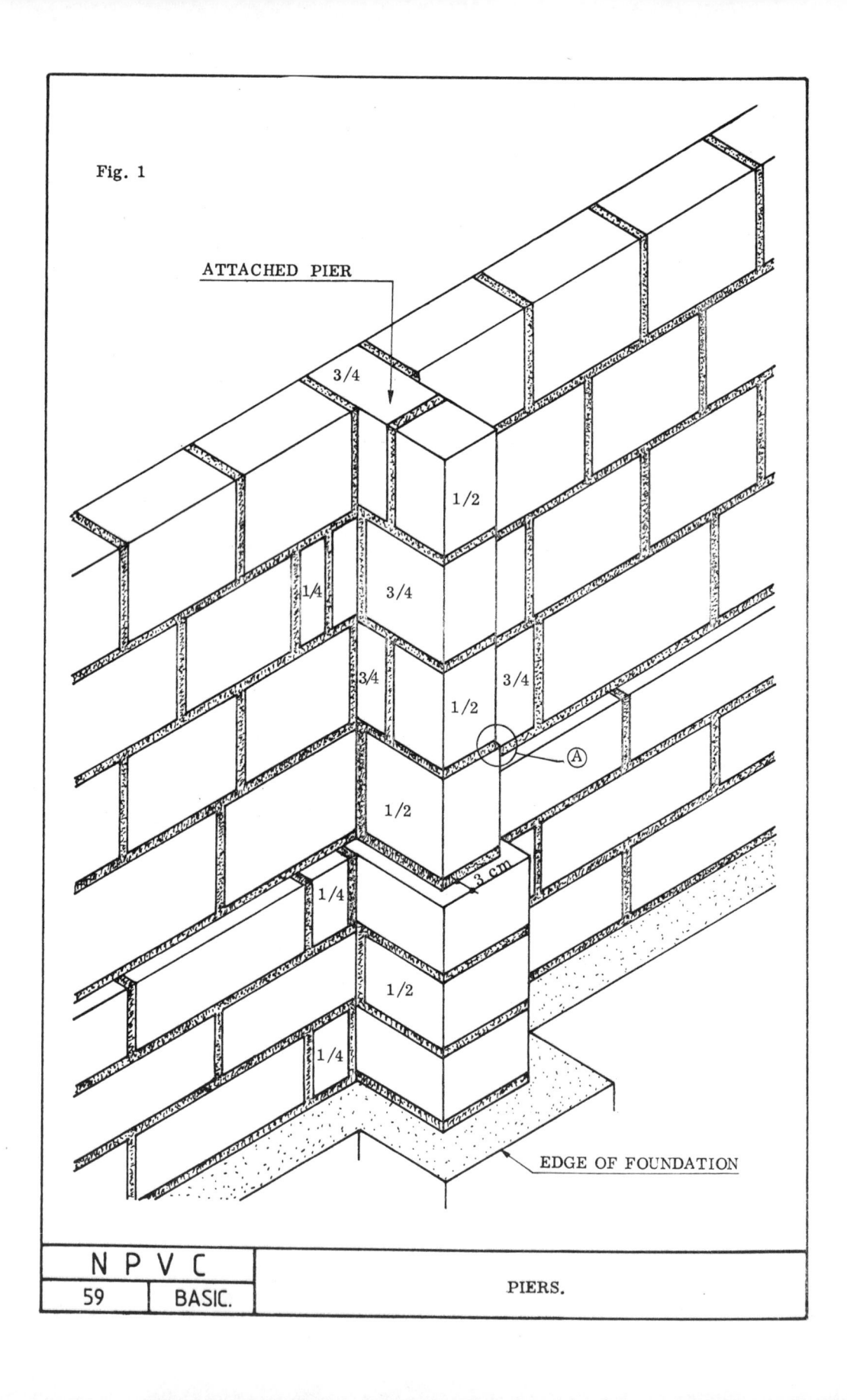

Fig. 1

ATTACHED PIER

3/4
1/2
1/4
3/4
3/4
3/4
1/2
1/2
Ⓐ
3 cm
1/4
1/2
1/4

EDGE OF FOUNDATION

N P V C		
59	BASIC.	PIERS.

PIERS

ATTACHED PIERS

Attached piers; also called engaged piers, wall piers, blind piers or pilasters, are piers partly sunk into a wall and properly bonded into it.

Normally the visible part of a pier projects only slightly from the wall, but in Rural Building the projection may be as much as the thickness of the wall or even more.

Formerly, attached piers were most often used as decorative elements. The Rural Builder, however, uses attached piers chiefly to strengthen walls. At the same time he saves valuable materials such as cement, reinforcement bars and timber for formwork that would be needed for a reinforced concrete pillar.

The construction of an attached pier is very similiar to that of a T-junction. The only difference is that the wall which joins the front wall is very short, and with a stopped end, thus forming an attached pier (Fig. 1).

The illustration shows that the bonds used to construct this pier are the same as those introduced in the chapter on T-junction bonds. Compare it with the text and illustrations from pages 38 to 44.

- NOTE: The plinth course is set back everywhere from the footings by 4 cm, except on the back side of the attached pier where it is set back only 3 cm (Fig. 1).

Do not forget to build the landcrete blocks flush with the plumbed (good) face of the plinth course. This means that on the other face of the wall there will be a 1 cm set-back caused by the 1 cm difference in size between the sandcrete blocks in the plinth course and the landcrete blocks in the rising wall (Fig. 1, point A).

NOTES:

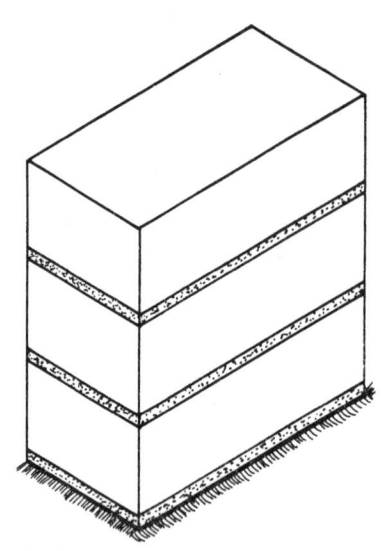

Fig. 1
SANDCRETE PIER

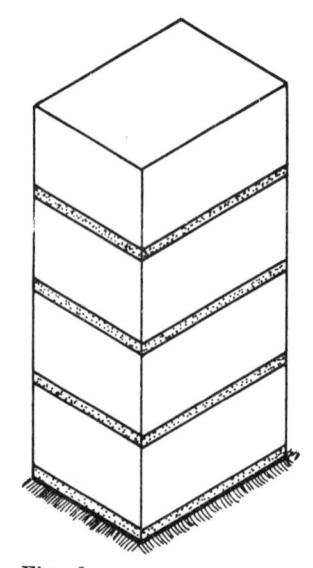

Fig. 2
LANDCRETE PIER

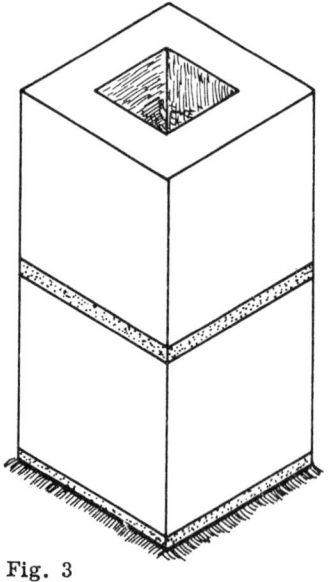

Fig. 3
PIER BUILT WITH OPEN BLOCKS

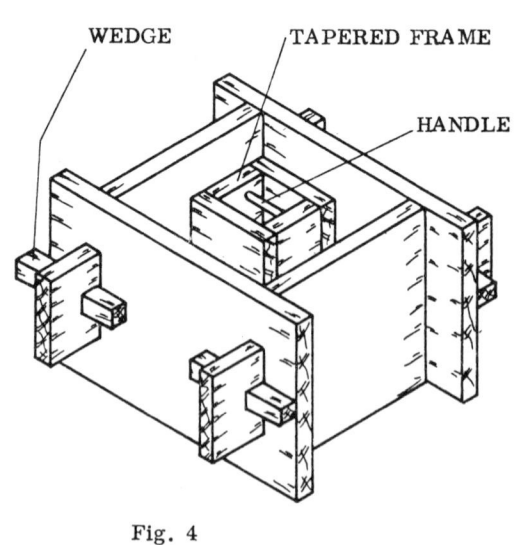

Fig. 4
WOODEN MOULD

WEDGE · TAPERED FRAME · HANDLE

NPVC		PIERS.
61	BASIC.	

SIMPLE PIERS

A pier is a pillar-shape of brickwork, blockwork or stone which usually has a square or rectangular section, and supports a load. In Rural Building its mass also helps to anchor the roof structure.

The easiest way to construct a pier is by simply laying full blocks flatwise one above another (Figs. 1 & 2). This will be sufficient in situations where the pier does not have to carry a very heavy load.

The disadvantage of this construction is that any roof anchorage must be fixed to the sides of the pier.

The best way to anchor a member of the structure to a pier is through the centre of the pier.

So-called perforated blocks can be made by using a specially made wooden mould (Fig. 3). In contrast to the common type of perforated block which has many smaller holes, this type has only one large hole through its centre. It is also called an open block.

Open blocks are made by casting mortar between two frames (Fig. 4). The bigger frame is made similarly to the wooden mould described in the Reference Book, Tools section, page 29. The smaller frame is tapered and has a handle to make it easier to remove from the block.

After the pier is built, the anchoring bar is inserted in the hole and the remaining space is filled up with mortar or concrete.

NOTES:

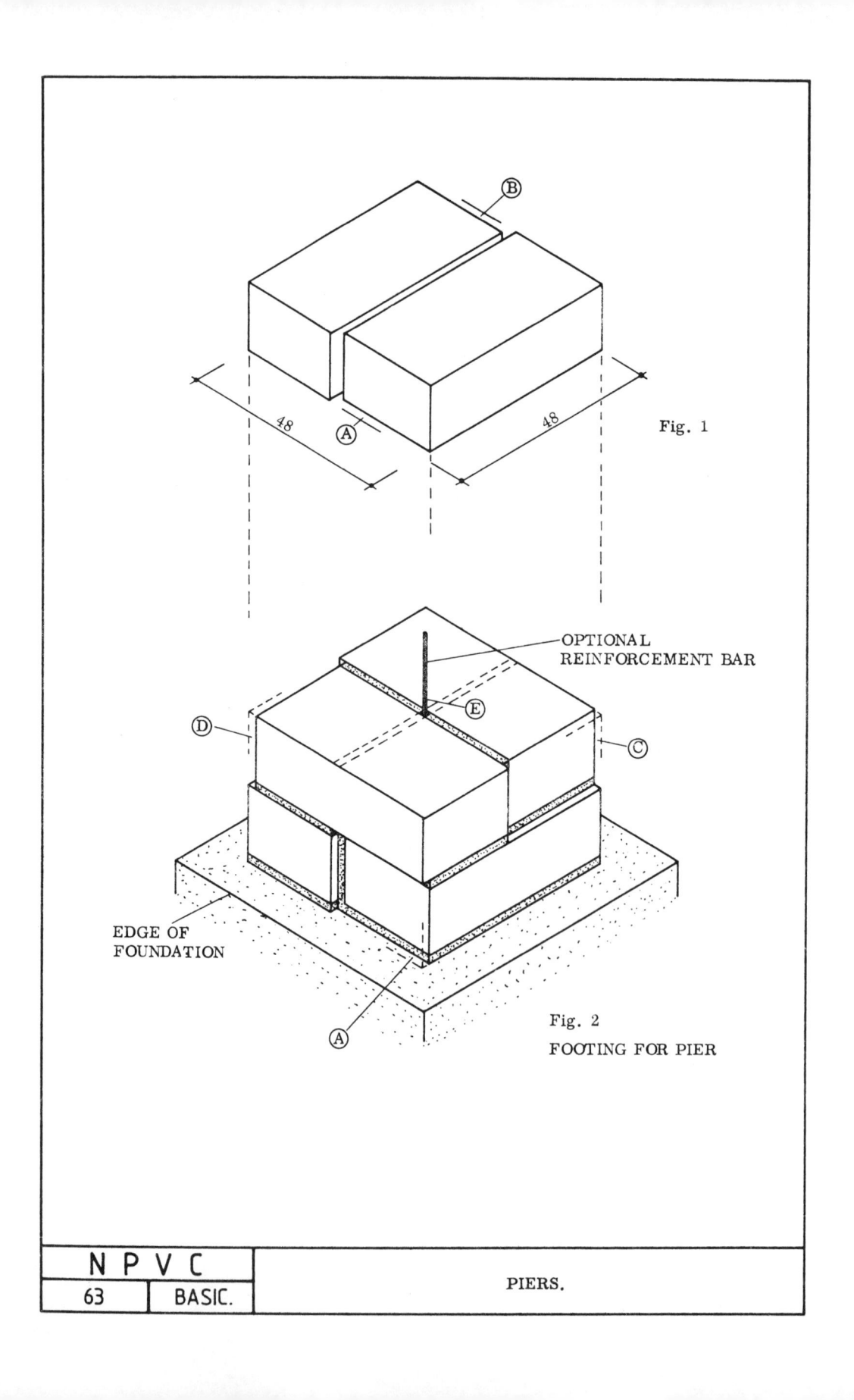

Fig. 1

OPTIONAL REINFORCEMENT BAR

EDGE OF FOUNDATION

Fig. 2
FOOTING FOR PIER

N P V C		
63	BASIC.	PIERS.

FOOTINGS FOR PIERS

There are situations where piers are needed which not only have to carry heavy loads but also have to be very heavy in themselves. This is to anchor the roof against the suction of strong winds.

In order to save valuable building materials such as cement, reinforcement bars and timber, as well as to reduce the construction time, blockwork piers are often built.

There are several possible bonding arrangements. The bonds introduced in this chapter represent only a few types, but they will meet the requirements of the Rural Builder.

The type of bond which is used depends largely on which materials are available and on the size of pier which is desired.

The opposite illustrations show a bond for footings with sandcrete blocks laid flatwise. Refer to page 67 for another footing method. Both bonds can also be used to build up an entire pier, in case a heavy duty pier is desired.

All the courses in the illustration on the left consist of a pair of blocks which are set 2 cm out of line with each other at the header sides. You can see this 2 cm difference at points A, B, C, D in Figs. 1 and 2.

This is done to maintain a square shape of 48 cm by 48 cm, since the width of the two blocks plus a joint is 48 cm, while the length of each block is only 46 cm. Since the courses cross each other, a reinforcement bar for anchorage could be built-in between the cross joints (Fig. 2, E).

NOTES:

Fig. 1
SANDCRETE PIER EDGEWISE

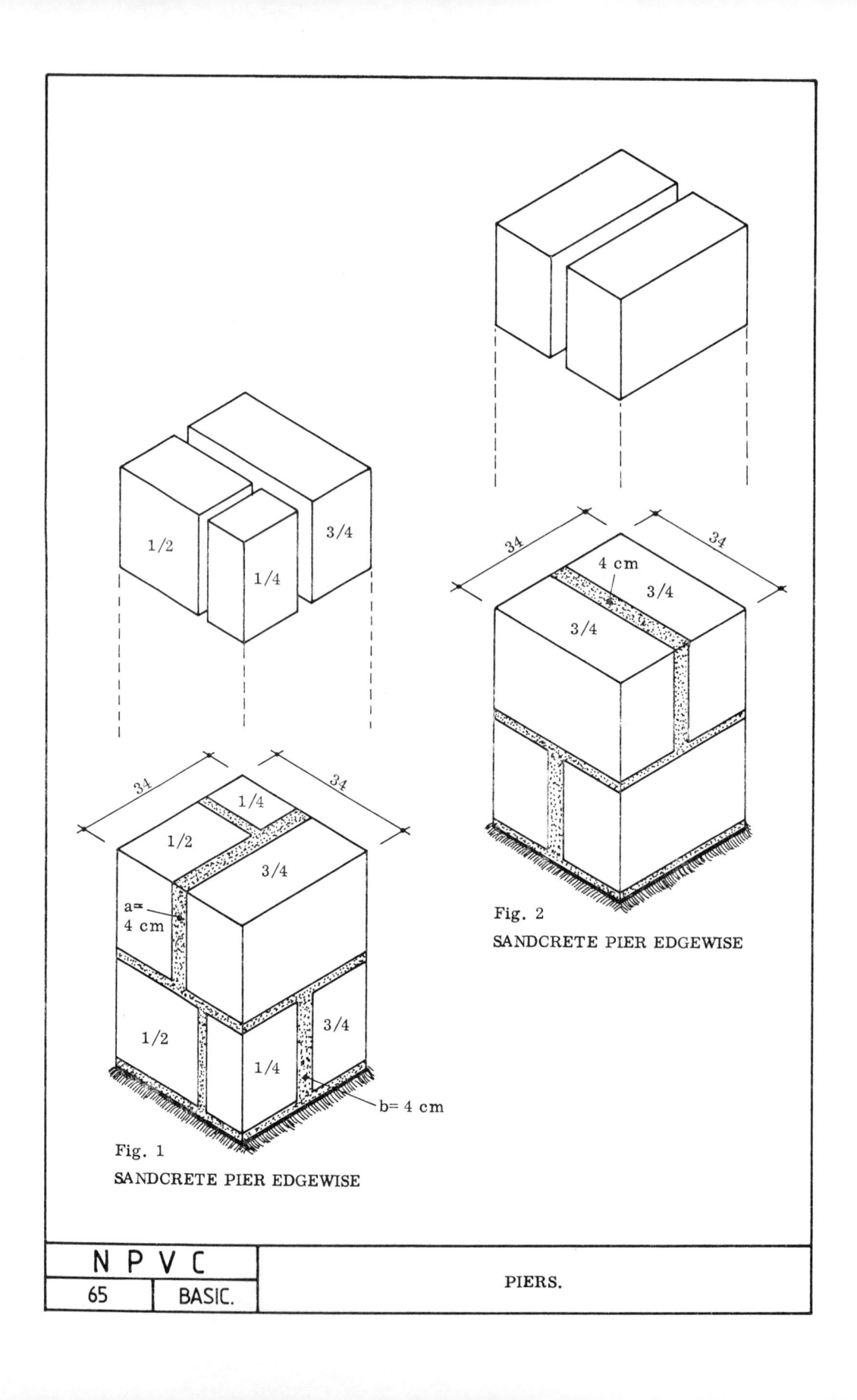

Fig. 2
SANDCRETE PIER EDGEWISE

N P V C		
65	BASIC.	PIERS.

BONDS FOR SANDCRETE PIERS

The following two bonds are also square-shaped but have shorter dimensions than the footing bonds: 34 cm by 34 cm.

Fig. 1 shows the best bond for a pier with the above measurements. This is because each course uses a total of $1\frac{1}{2}$ blocks, which means that there is no waste.

The through-going cross joints in each course must be widened (Fig. 1, a & b) to 4 cm in order to obtain a width of 34 cm across the two blocks; 34 cm is the length of the 3/4 blocks.

The 3/4 block in each course is set over the 1/4 block below, and across the 4 cm joint, making an alternating arrangement as shown (Fig. 1).

Fig. 2 shows another bond for a sandcrete pier. The pier is built entirely of 3/4 blocks. The cross joints all go through and all have a 4 cm thickness.

It is obvious that this type of pier should be built only if there are a lot of 3/4 blocks left-over from another construction, because otherwise all the 1/4 blocks are left-over and wasted.

NOTES:

3/4
3/4
1/4
1/4

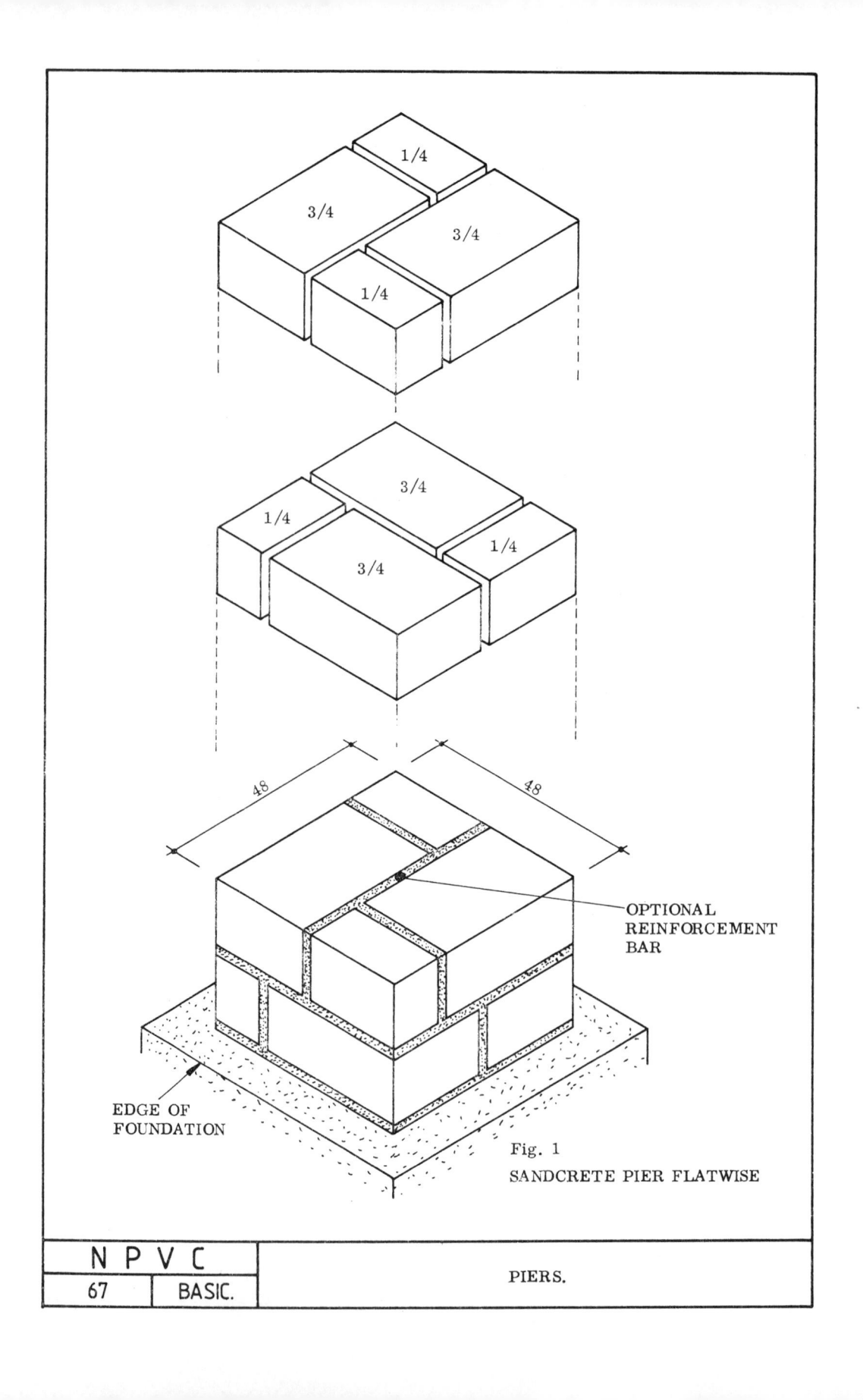

3/4
1/4
3/4
1/4

48
48

OPTIONAL
REINFORCEMENT
BAR

EDGE OF
FOUNDATION

Fig. 1
SANDCRETE PIER FLATWISE

N P V C		
67	BASIC.	PIERS.

REINFORCEMENT BAR

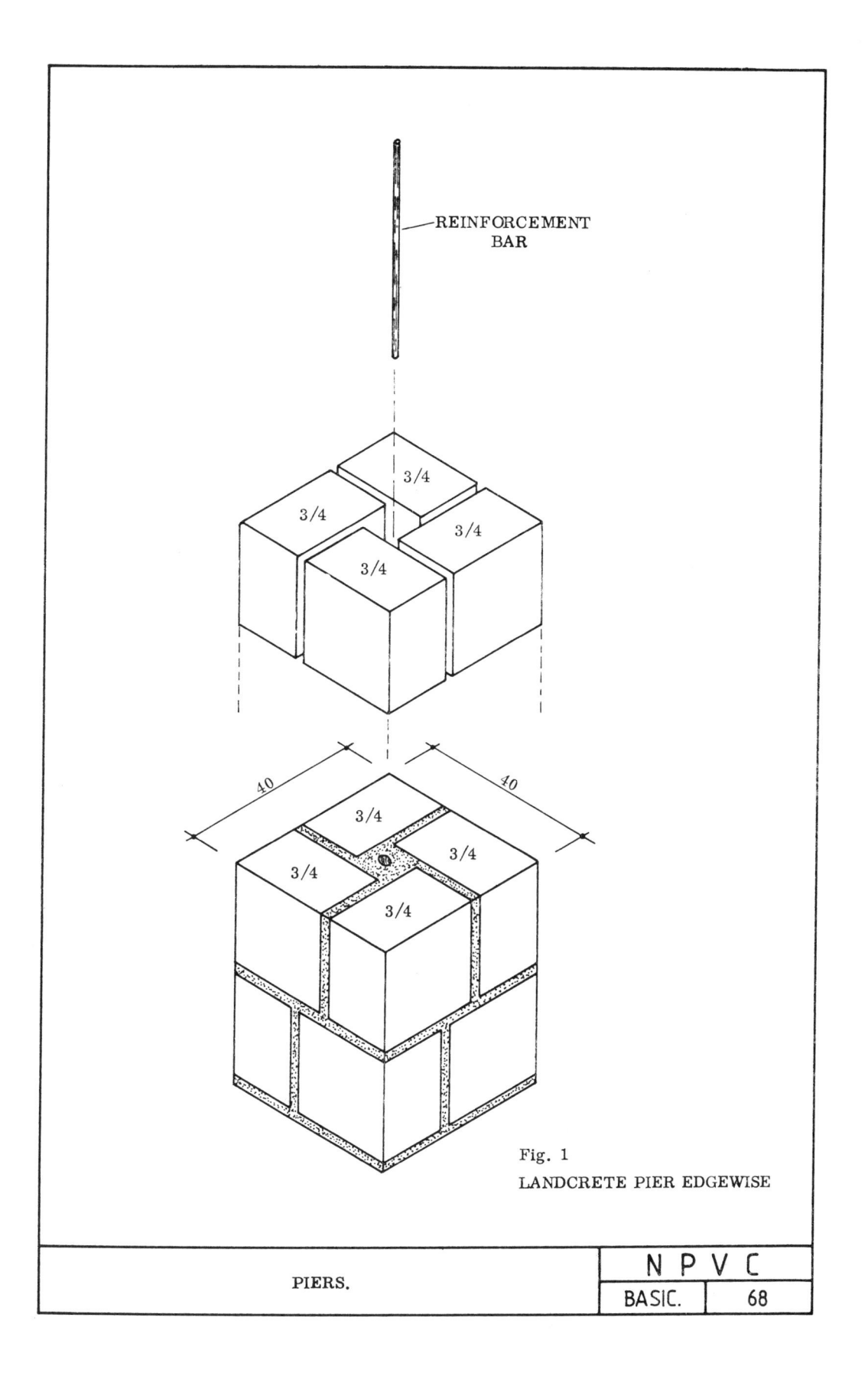

3/4
3/4
3/4
3/4

40
40

3/4
3/4
3/4
3/4

Fig. 1
LANDCRETE PIER EDGEWISE

PIERS.

N P V C
BASIC. 68

Fig. 2
LANDCRETE PIER EDGEWISE

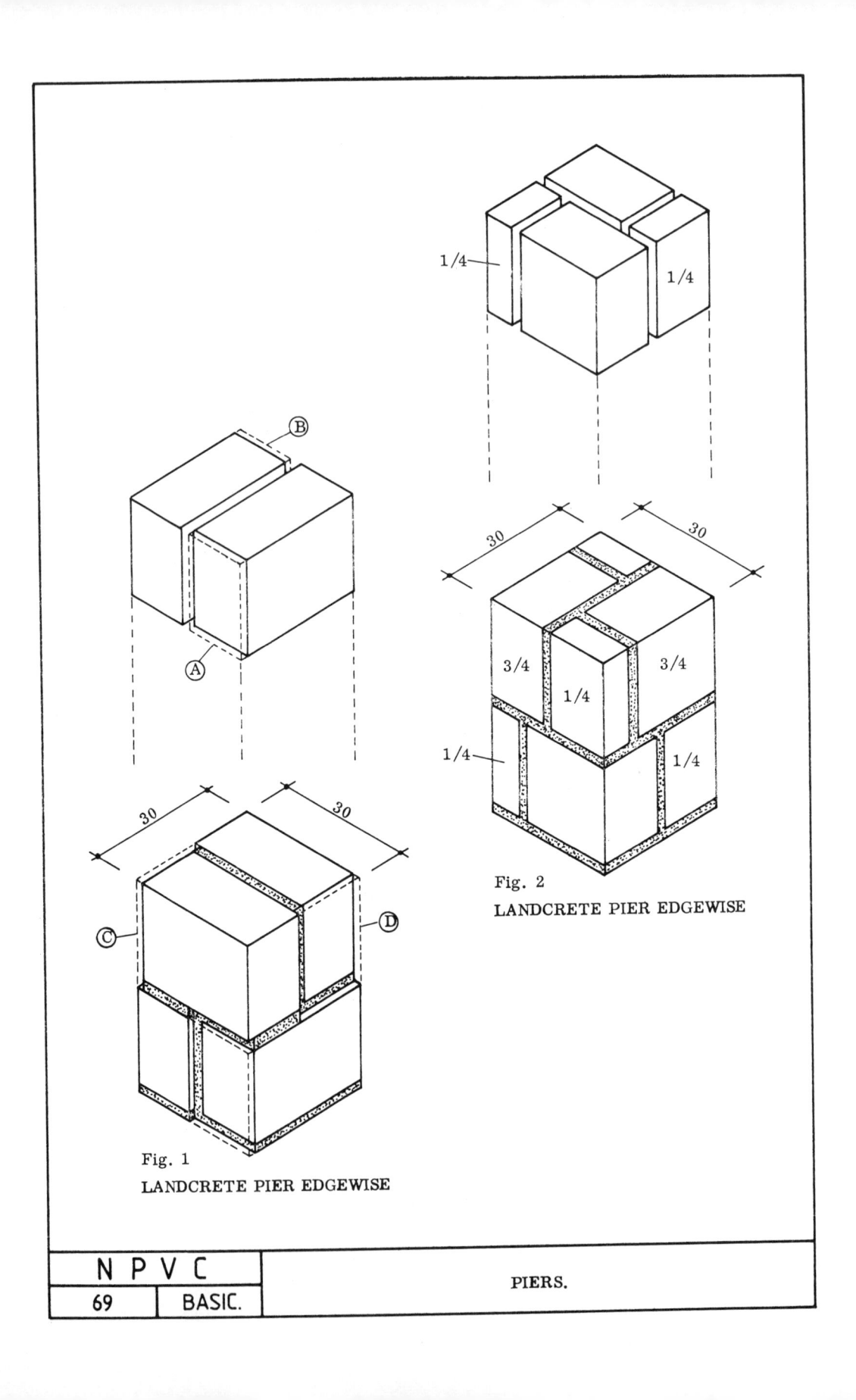

Fig. 1
LANDCRETE PIER EDGEWISE

N P V C		PIERS.
69	BASIC.	

BONDS FOR LANDCRETE PIERS

The examples of bonds for landcrete piers shown on pages 68 and 69 are similar to some of the bonds already explained. Fig. 1 shows a bond which is basically the same as the footing bond on page 63, except the blocks are laid edgewise.

Each course consists of two full blocks laid edgewise and across the course below. In order to maintain a square shape of 30 cm by 30 cm, the blocks are set 1 cm out of line with each other at their header sides (see points A, B, C, & D).

The bonding method seen in Fig. 2 requires cutting two blocks per course and results in a rather weak bond. The Rural Builder should avoid making this type of pier except when he needs to use up some part-blocks left over from another construction.

Fig. 1 on page 68 shows another bonding method for landcrete piers with the blocks laid edgewise. This method has the advantage that an iron rod can be set into the centre of the blocks, either to reinforce the pier or to act as an anchor, or both.

- NOTE: Landcrete piers should not carry heavy loads such as a truss or a concrete beam, because they are too weak. They may be used to support and anchor overhanging rafters of a roof above a verandah.

If a landcrete pier has to carry heavy loads, its dimensions should be increased to a minimum of 55 cm by 55 cm.

NOTES:

Fig. 1

Fig. 2 — END GRAIN

Fig. 3

Fig. 4

Fig. 5

Fig. 6

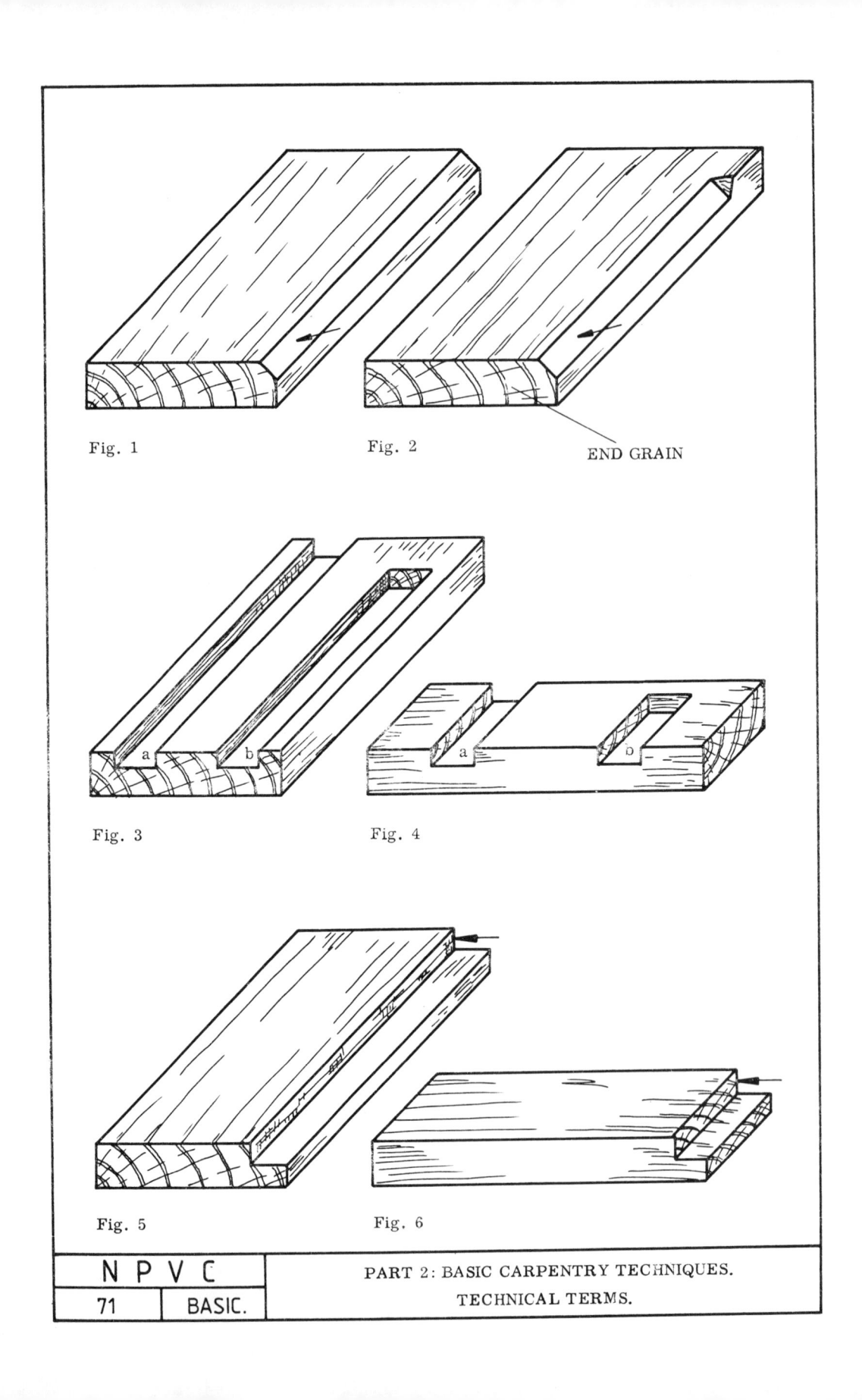

NPVC	PART 2: BASIC CARPENTRY TECHNIQUES.
71 BASIC.	TECHNICAL TERMS.

PART 2: BASIC CARPENTRY TECHNIQUES

TECHNICAL TERMS

- GRAIN: This refers to the direction of the wood fibres. Length is measured along the direction of the grain. Width is measured across the grain at right angles to the length. When wood is cut across the grain, END GRAIN is exposed.

- WITH THE GRAIN: This term is used in connection with planing. If the fibres are cut cleanly and smoothed down by the cutting iron, the wood is said to be planed with the grain; like stroking a dog's coat so the hair lies down smoothly.

- AGAINST THE GRAIN: This means that the plane goes in the opposite direction, lifting and breaking the wood fibres and leaving a rough surface; as if a dog's coat were brushed the wrong way and roughened.

- STRAIGHT GRAIN: The wood fibres lie straight and parallel to the length of the piece of wood. Such wood planes smoothly and easily.

- CROSS GRAIN: The wood fibres do not lie parallel to the length of the piece. This makes the wood hard to work.

- BEVEL: This is made by planing off the sharp edge to form a new surface which is not at right angles to the side of the piece of wood. A CHAMFER is a special bevel, cut at 45 degrees. A "through" chamfer or bevel runs the whole length of the edge (Fig. 1). A "stopped" chamfer or bevel is stopped at one or both ends (Fig. 2).

- GROOVE: This is a recess cut along the grain. A "through" groove runs the whole length of the piece (Fig. 3, a); while a "stopped" groove is stopped at one or both ends (Fig. 3, b).

- TRENCH: This is a recess cut along the grain. A trench can also be either through (Fig. 4, a) or stopped (Fig. 4, b).

- REBATE: This is a recess cut along the edge or across the end of a board as in Figs. 5 & 6.

- TRUE: In woodworking this indicates that a surface is flat and perfectly level.

- SQUARE: Square angles are exact 90 degree angles. "Square" is used to describe pieces in which all the corners and edges have 90 degree angles.

- SHOULDER: The vertical portion of a trench or rebate (arrows, Figs. 5 & 6).

Fig. 1

Fig. 2

Fig. 3

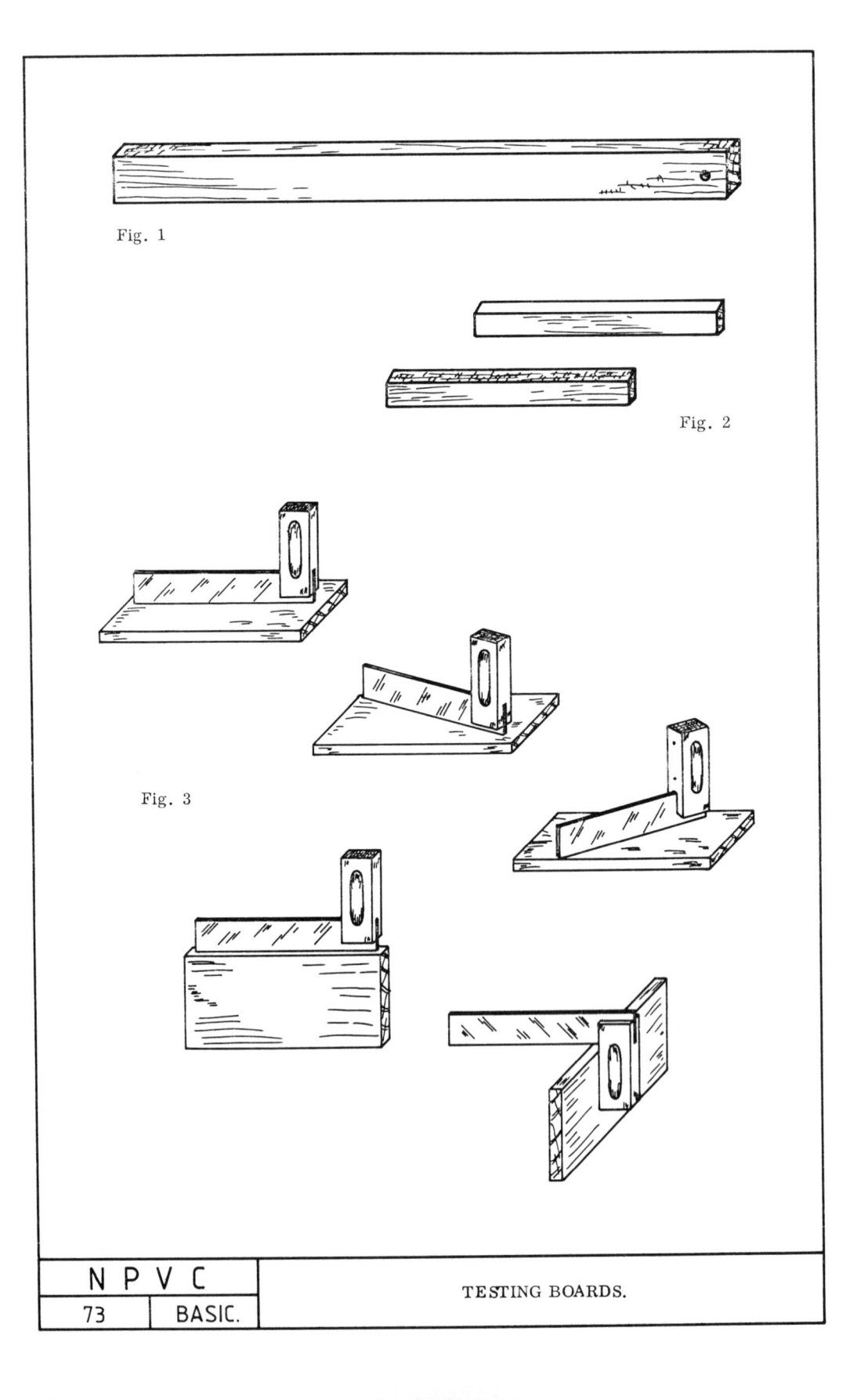

N P V C		TESTING BOARDS.
73	BASIC.	

TESTING BOARDS

When you prepare a board for use in some project, you must make certain tests on it to make sure that it is flat and true in all directions and that the angles and corners are all square. These tests are made during the actual preparation of the timber, but we describe them here separately because they are generally useful techniques which you will need again and again in your work.

Before you continue reading, look in your Rural Building Reference Book, Tools section, page 36, and read about the try square, which is one of the tools you will need for testing boards. You will also need winding strips (Fig. 2) and a straight edge (Fig. 1), which is usually a piece of wood with one long edge that you are sure is perfectly flat and straight.

Also look in the Reference Book, Materials section, page 132 and find out what is meant by the words: twisting; cupping; and bowing.

WINDING STRIPS

These are used as an aid to help you to see if a board twists or "winds". They are two strips of wood about 35 cm long, 2,5 cm wide and 1,5 cm thick. The top edge has a bevel and all the edges must be perfectly straight. One of the two strips may be made darker so that sighting along them is easier (Fig. 2).

HOW TO TEST SMALL WORK PIECES

Test with the try square or the edge of a jack plane in different positions for flatness. Also test the squareness of the edges with a try square at a few different spots (Fig. 3).

NOTES:

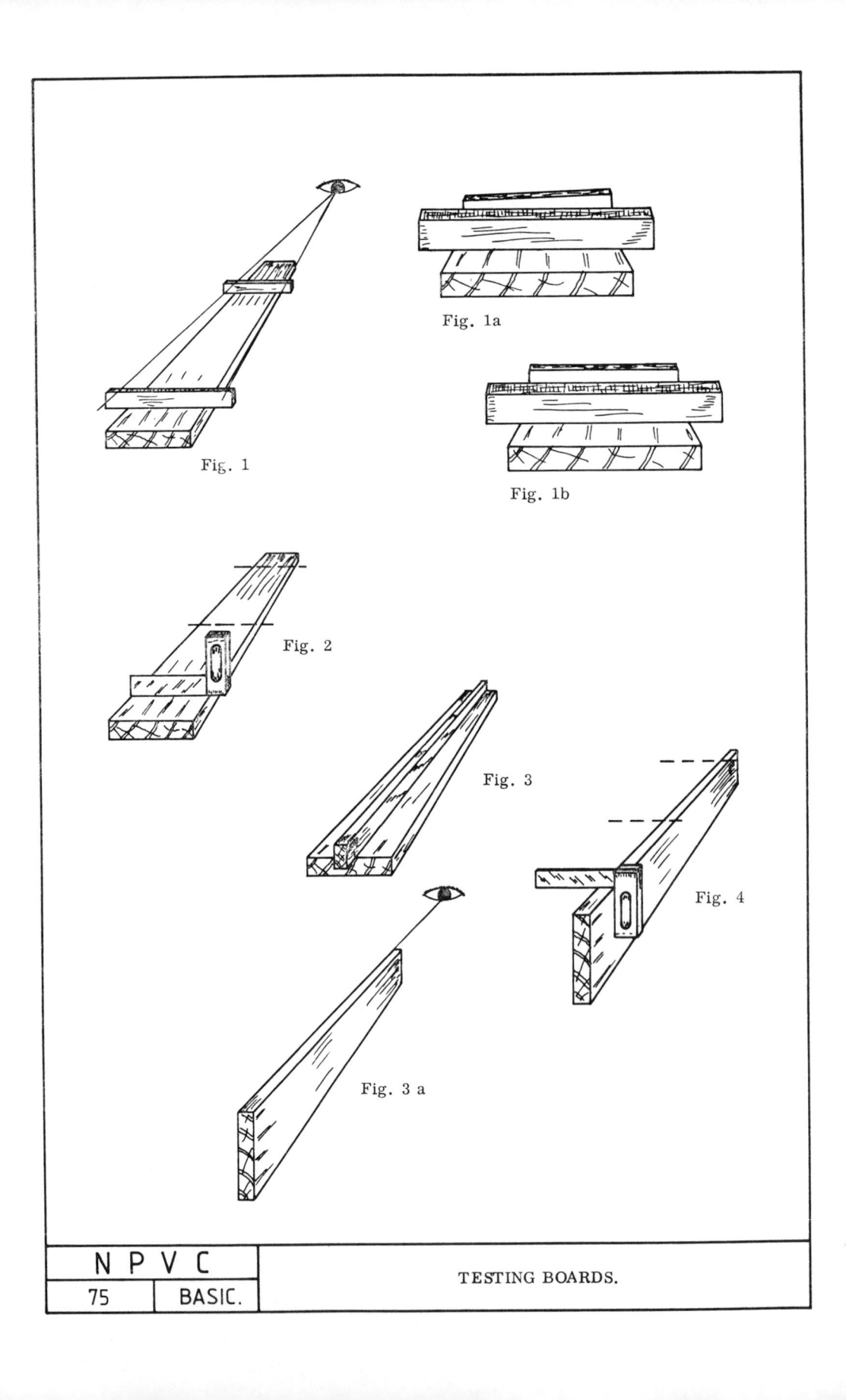

Fig. 1

Fig. 1a

Fig. 1b

Fig. 2

Fig. 3

Fig. 3 a

Fig. 4

N P V C	TESTING BOARDS.
75	BASIC.

HOW TO TEST LONG WORK PIECES

Test for winding (twisting) using the winding strips; one at each end of the board, across the grain. Go to the end of the board and look along the board, with your eye just at the same level as the strips (Fig. 1). When the top edges of the strips do not appear parallel (Fig. 1a), the board is not flat. Check with the winding strips at different spots, making sure that the strips are parallel (Fig. 1b).

Test for cupping by putting a try square across the grain at different spots along the board. If you check this against the light, you will see all of the uneven places (Fig. 2).

To check for bowing you can do the same test, using a straight edge along the grain (Fig. 3).

For long boards you can sight along the boards with one eye closed, to see the places which are uneven (Fig. 3a).

Using a try square, check whether the angle between the face side and the edge is exactly 90 degrees. Make this test at several places (Fig. 4).

When all these tests have been performed with satisfactory results, the board will be straight and true in all directions and it is ready to be used in a project.

NOTES:

TESTING BOARDS.

N P V C
BASIC. 76

Fig. 1

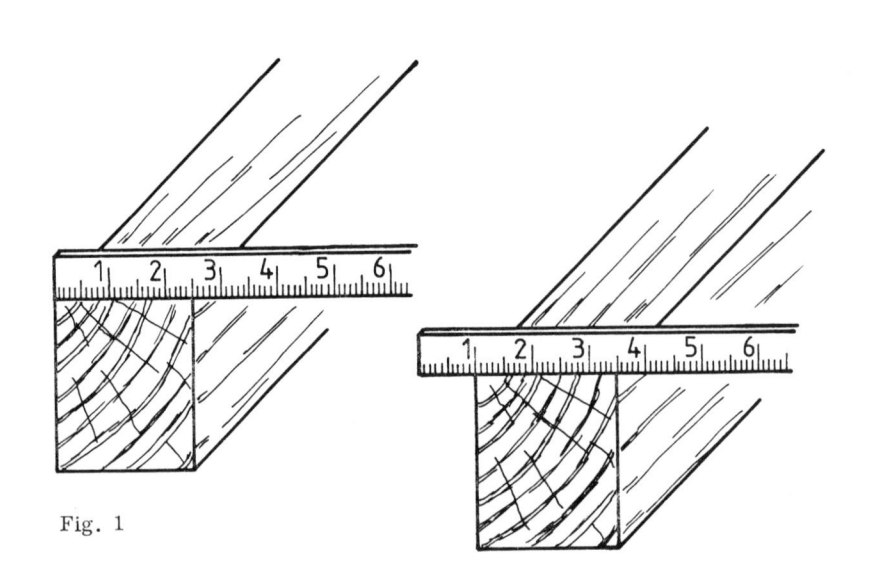

Fig. 2

Fig. 3

MEASURING AND MARKING OUT TIMBER.

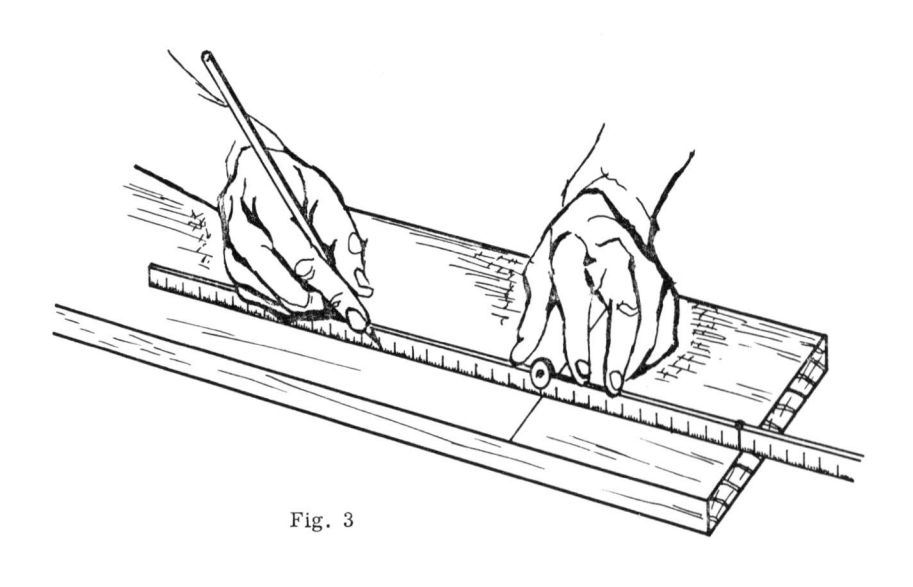

MEASURING AND MARKING OUT TIMBER

Accurate measuring and marking out are the first requirements for success in the building trade. Common measuring and marking tools are the folding rule and zig-zag rule (Reference Book, page 11).

MEASURING WITH A RULE

To measure between two points, place the rule on one point and read the mark nearest to the other point (Fig. 1).

When the end of the rule has become worn and inexact, you can still get an exact measurement. Place the 1 cm mark of the ruler at the first point and read at the second point. The true measurement is that reading minus 1 cm. For example, the measurement in Fig. 2 is: 3,5 cm minus 1 cm equals 2,5 cm.

MARKING OUT WITH A RULE

To mark out measurements with a rule, place the end of the rule (or the 1 cm mark) carefully at the start of the measurement and then make a fine mark with a pencil exactly even with the marking on the rule at the correct distance.

For very accurate marking and measuring, lay the rule on its edge so that the marks on the rule touch the work (Fig. 3).

To mark out several measurements on a line, it is best to mark all of the measurements without raising the rule or moving it. If the rule is moved and each measurement is made separately, there is a much greater possibility of error.

NOTES:

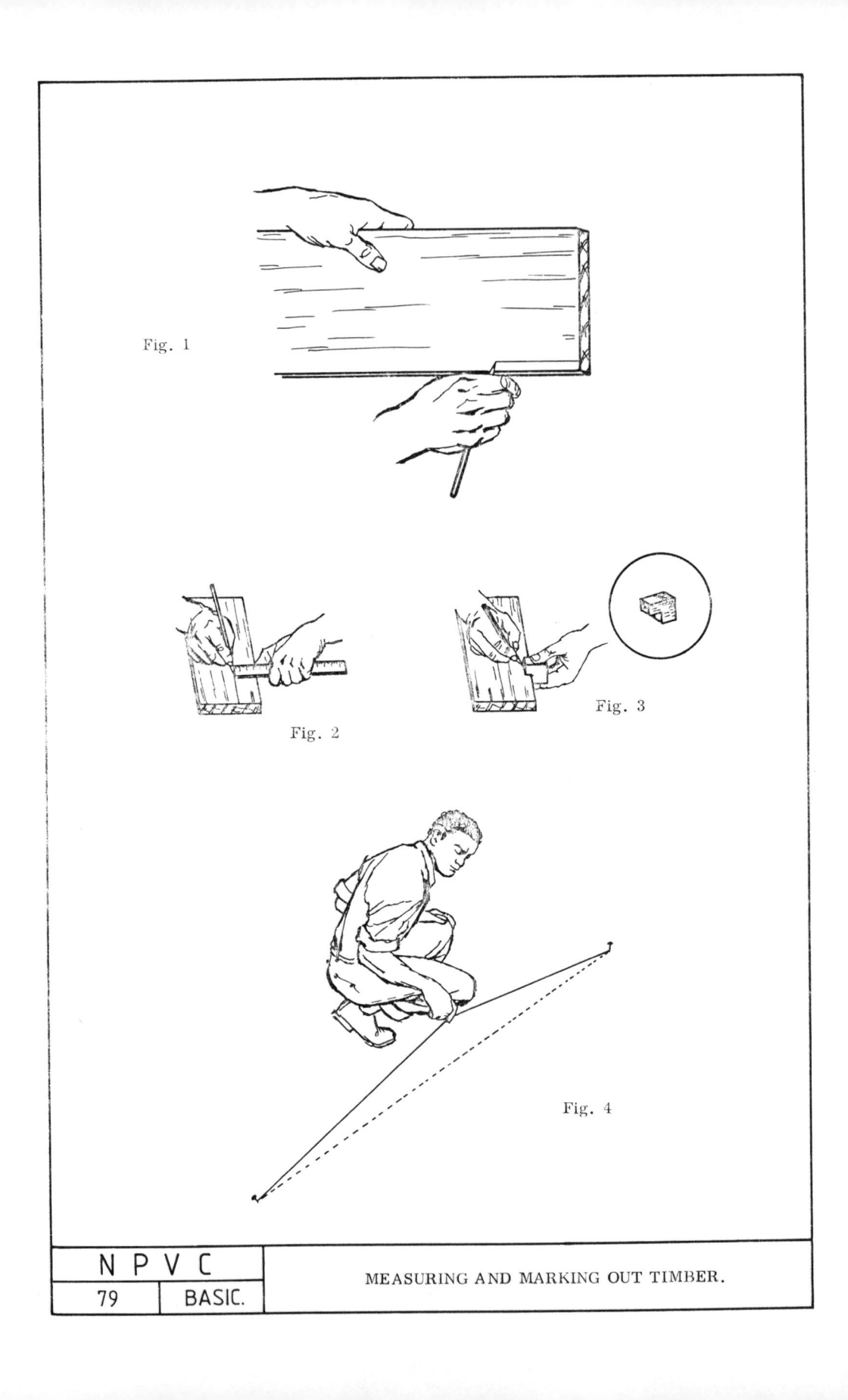

Fig. 1

Fig. 2

Fig. 3

Fig. 4

MEASURING AND MARKING OUT TIMBER.

NPVC 79 BASIC.

MARKING WITH A PENCIL

When a marking gauge (Reference Book, Tools, page 39) is not available, straight lines can be gauged along timber by one of the following methods.

- Grasp the pencil lightly in your closed fist with the point protruding the desired distance. For example, to make a line 1 cm from the edge of the timber, the point of the pencil should stick out exactly 1 cm from your fist. Now pull the pencil along the board keeping your thumbnail pressed firmly to the edge of the board (Fig. 1).

- To gauge lines further from the edge of the board, use a rule and a pencil. Grasp the rule in one hand with your thumbnail at the desired marking. Then draw the rule along while keeping your thumbnail against the edge of the board. With your other hand hold a pencil at the end of the rule to make the line (Fig. 2).

With these methods you should be careful not to get splinters in your fingers.

- Another method of drawing lines parallel to an edge is with a small pencil gauge. This is simply a small, rebated wooden block which is pressed against the edge of the timber and used to guide the pencil, as shown in Fig. 3. The pencil gauge is often used to mark out the position of chamfers.

- A straight edge can also be used in marking out longer lines.

MARKING WITH A CHALK LINE

A quick and simple way of marking out a straight line on any surface is with a chalk line. This is simply a piece of string that has been rubbed with chalk until it is coated in chalk dust (charcoal may also be used for this purpose).

To use the chalk line, stretch the line between two points which are the ends of the line you want to mark. Hold it in place by tying the ends to nails, or have a helper hold it for you.

Lift the line up in the middle and allow it to snap back (Fig. 4), making a straight chalk line on the surface.

NOTES:

MEASURING AND MARKING OUT TIMBER.

TIMBER CONSTRUCTIONS

The last few lessons were about the very basic things you will need to know before you can actually make any construction of wood. Therefore we started with some technical terms which are important because you will see them again and again in the lessons and will use them in practical work.

The next sections include all the information you will need to prepare for a project: how to get the wood pieces to the right size and shape; how to mark the timber; and how to plan a project using a cutting list. Included here is a section on nailing and one on driving screws. Before you study those sections, look up nails and screws in the Reference Book, Products, pages 207 to 211, so that you are familiar with the parts of nails and screws and how they look.

Each section starts with a list of tools. Refer to the Reference Book, Tools section, and learn about each of these tools before you study the rest of the section.

NOTES:

PREPARATION OF TIMBER

In order to carry out the construction of any practical exercise or project, it is essential to have the wood pieces for the project prepared so that they have the correct size, true and flat surfaces, and square corners and angles.

There is of course a correct procedure to be followed for this preparation. It is important to carry out the following steps in their correct order on every piece of timber, whether large or small.

An easy way to remember the steps in their correct order is to keep the following word in mind: F E W T E L.

Face	Edge	Width	Thickness	End	Length
F	E	W	T	E	L

The tools which are required for this preparation are the bench, jack plane, straight edge, winding strips, pencil, try square, marking gauge, charcoal line, back saw, ruler and crosscut saw. Look these up in the Reference Book and make sure that you understand what they are and how they are used.

NOTES:

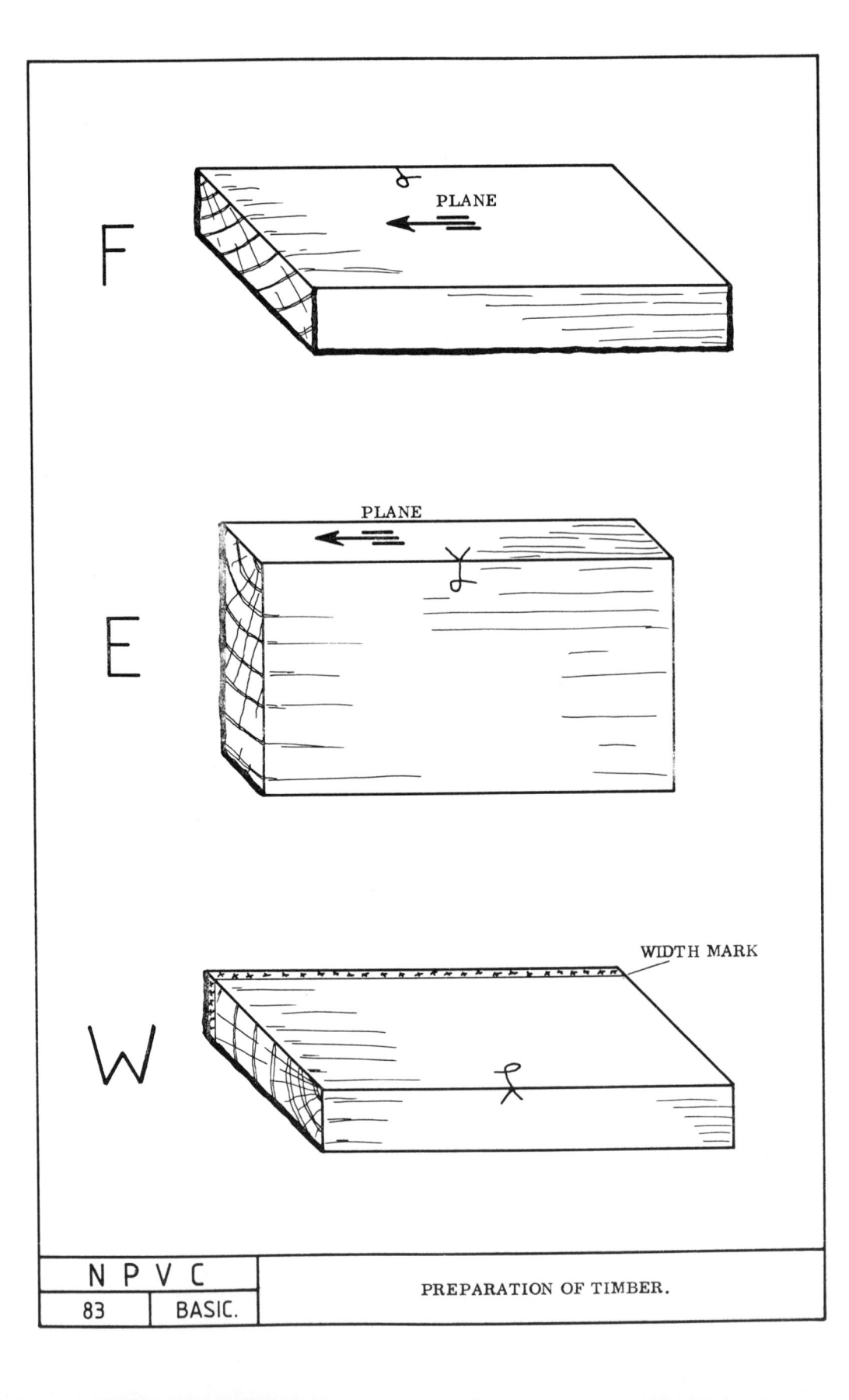

PREPARATION OF TIMBER.

F — PLANE

E — PLANE

W — WIDTH MARK

NPVC | 83 | BASIC.

SEQUENCE OF OPERATIONS FOR PREPARING TIMBER

Step 1. Plane the face side (F)

- Put your work on the bench with the better side up. If the board is not flat, put the hollow side down to keep it from rocking or moving around. In the case of very long or thin boards or twisted or deformed boards, put thin wedges of wood underneath where needed, to keep the work steady and keep it from bending in the middle during planing.

- Plane this side perfectly true.

- Test for flatness with the straight edge, winding strips and try square.

- Mark this side as the face side; the face mark should point to the edge which will be the face edge. The face edge will be the best edge of the board.

Step 2. Plane the face edge (E)

- Fasten your board to the side of the bench, with the face mark up. The planed side must be towards you and not against the bench, unless this would mean you would have to plane against the grain. In that case, you should turn the board so the other edge is up, keeping the face side towards you because the try square has to be set against the face side.

- Plane the face edge perfectly straight and square to the face side.

- Test for straightness with the straight edge and for squareness with the try square.

- Mark it with a face edge mark pointing to the face side.

Step 3. Plane the width (W)

- With a marking gauge, mark the width of the board. Press the gauge against the face edge and mark on the face side.

- Plane down to the middle of the gauge line and be careful to get a square, straight edge when you reach the mark.

- Test for flatness with a straight edge and try square.

NOTES:

PREPARATION OF TIMBER.

T

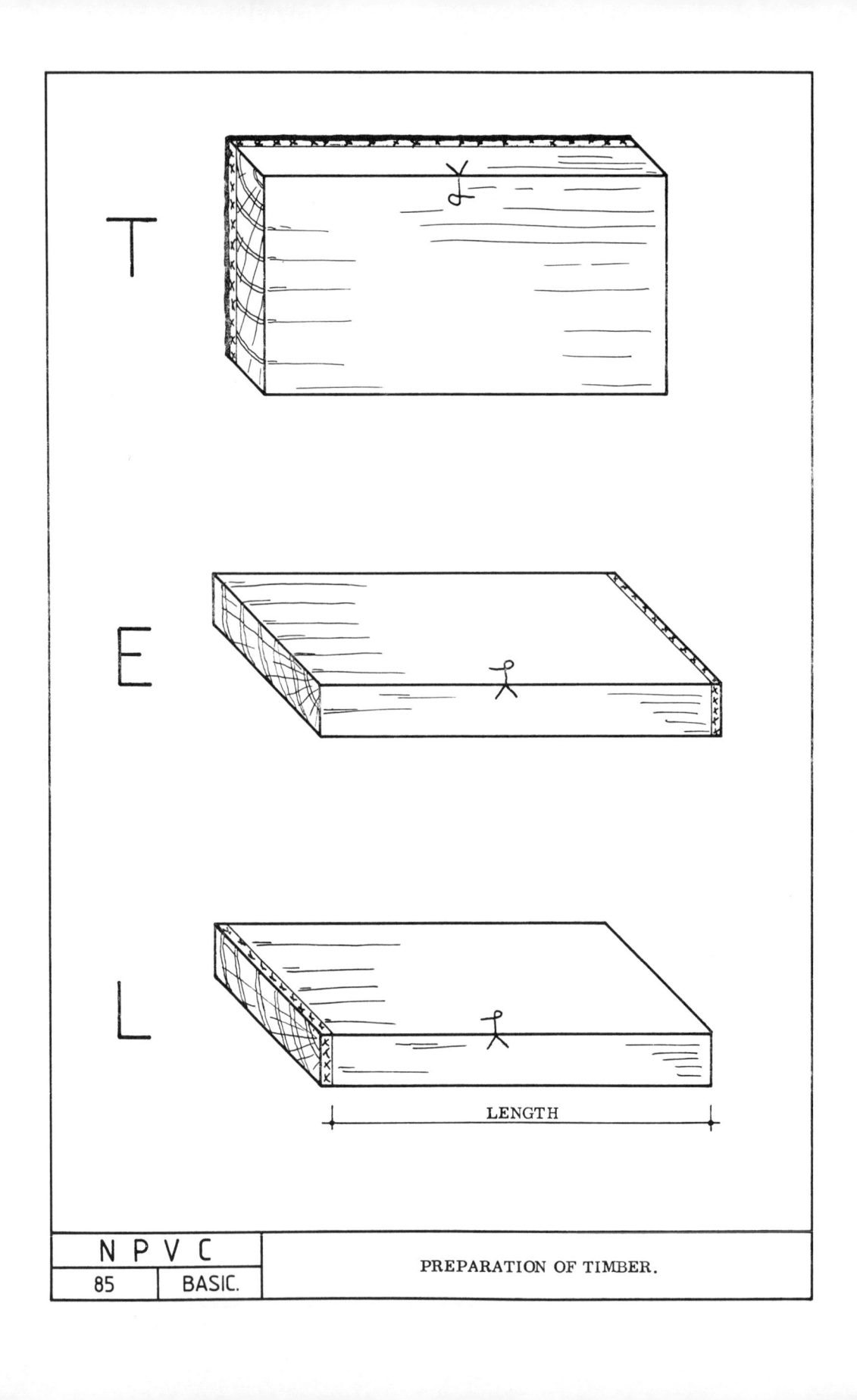

E

L

LENGTH

| NPVC | PREPARATION OF TIMBER. |
| 85 | BASIC. |

Step 4. Plane the thickness (T)

- Use a marking gauge to mark the thickness of your board. Press the gauge against the planed side and mark both edges.

- Plane down to the middle of the gauge lines.

- Test for flatness as you get near the two marked lines. Be sure to check for flatness across the board at several points. Check for winding with the winding strips.

Generally timber is not prepared to the exact length. Waste is left on each end to protect the corners from damage. If it is necessary to prepare it to the exact length, the procedure is as follows:

Step 5. Cut one end (E)

- With a try square, square the best end; with as little waste as possible.

- Cut that end perfectly square to the face side and face edge. Saw on the waste side of the line.

- Test for squareness in all directions.

- Mark the end with a cross, so that you know which is the prepared and tested end.

Step 6. Cut the length (L)

- Measure the required length from the prepared end and square with the try square.

- Cut that end perfectly square in all directions.

- Test it with a try square.

NOTES:

PREPARATION OF TIMBER.

Fig. 1

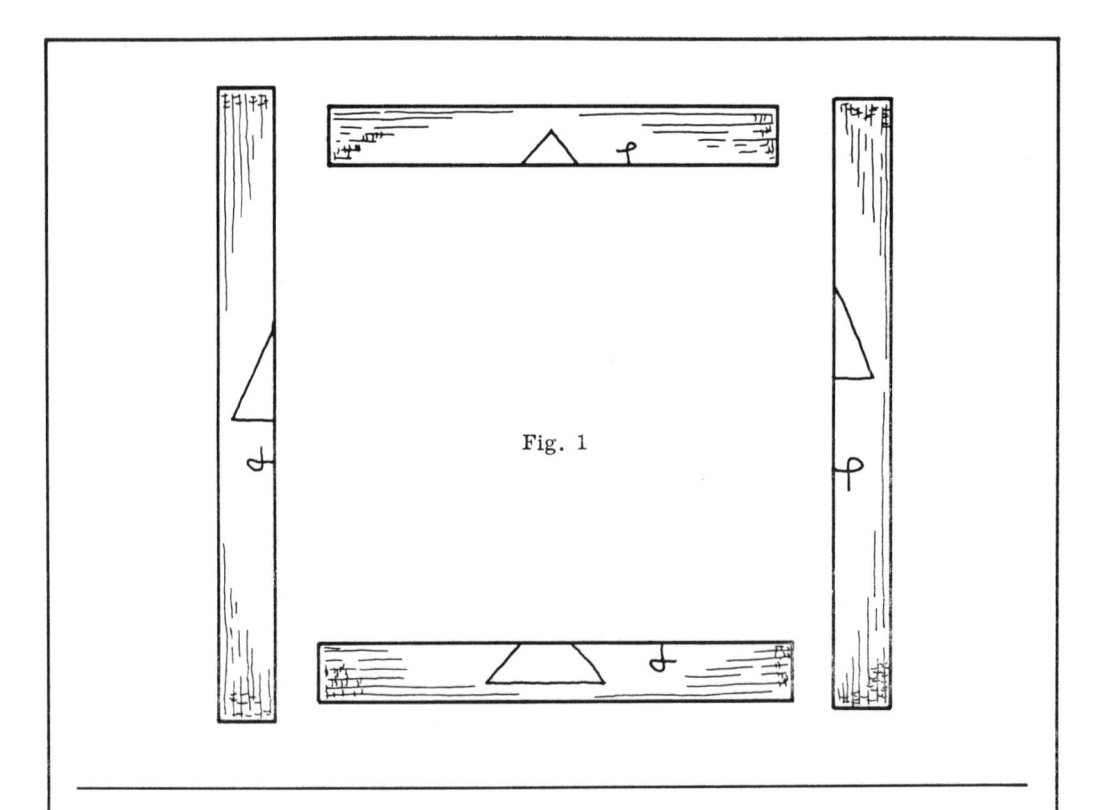

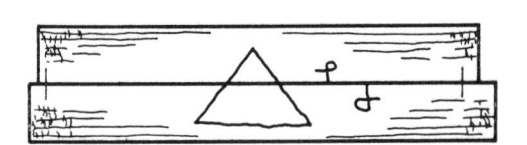

Fig. 2

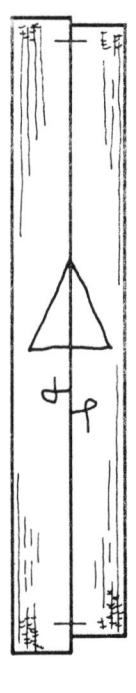

Fig. 3

PREPARATION OF TIMBER.

NPVC 87 BASIC.

TIMBER MARKS

The purpose of face marks, as they are shown under preparation of timber, is to show clearly the prepared and tested sides and the edges which are square.

During all further marking, squaring and gauging we should try to work from these sides and edges.

MARKING OF FRAMES

Making frames is an important part of Rural Building. We have to mark the members of a frame in a standard way so that we do not confuse their positions.

- Select and prepare the timber according to the sequence given in the section on preparation of timber. Lay out the various members as they will be when they are finally assembled into a frame (Fig. 1).

- Take care that all surfaces which will be visible in the finished product are of good timber and without defects. Usually these will be the face sides and face edges. Always try to keep the face sides on one side, and the face edges all on the inside or all on the outside.

- Now put the inside edges of the horizontal members together, mark the length of the members and put the triangular mark as shown in Fig. 2. The triangle always should point up.

- Now do the same thing for the vertical members. Put the inside edges together, mark the lengths and put the triangular sign, again pointing upwards (Fig. 3).

- All further marking should be done using these sides as a reference.

NOTES:

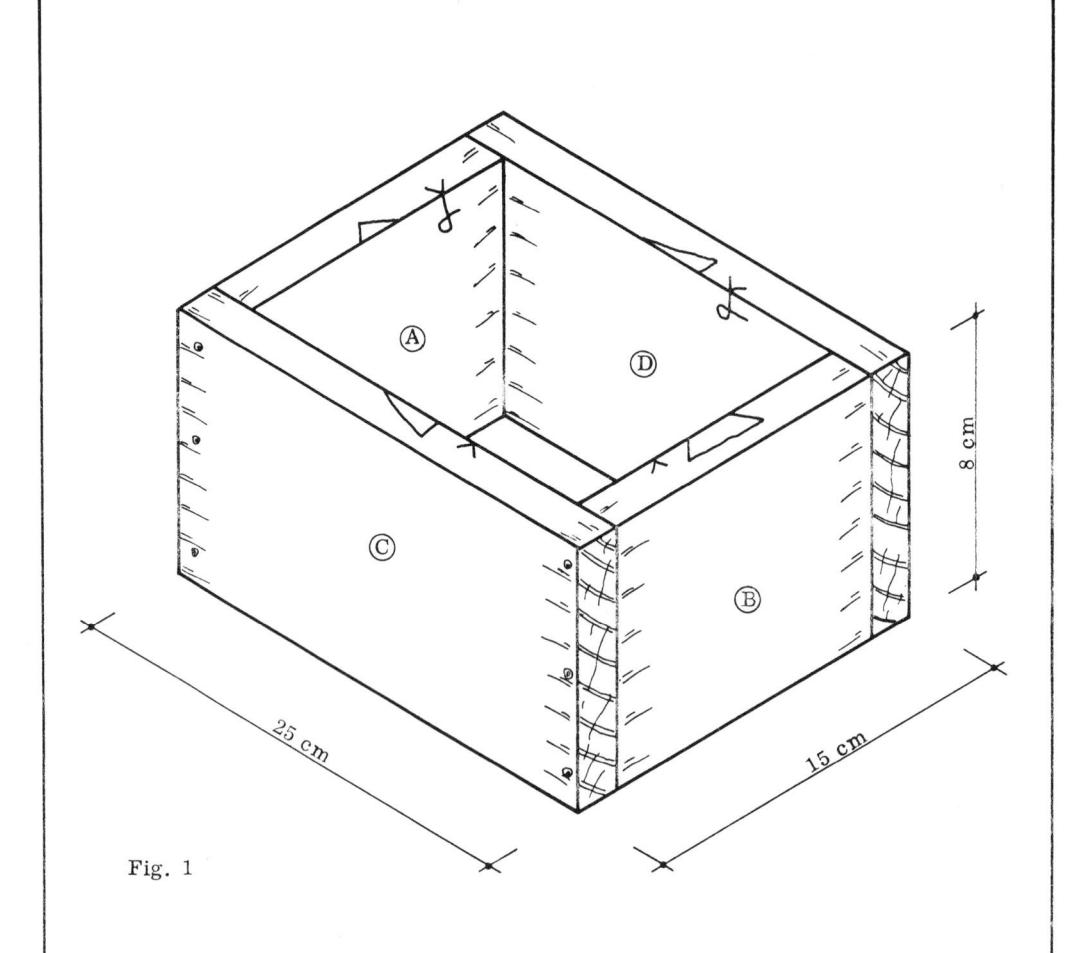

Fig. 1

NOTES:

N P V C		PREPARATION OF TIMBER.
89	BASIC.	

CUTTING LIST

After you design and make the drawing for a piece of work, you need to make a cutting list showing the length, width and thickness of all the parts. These will be the finished sizes. So that each part will be straight, true and smooth, we must begin with slightly bigger pieces to have an allowance for planing the sawn timber true.

The planing allowance for a board is:

3 mm extra in thickness

6 mm extra in width

12 mm extra in length

For square pieces the allowance is:

3 mm extra on each side

12 mm extra in length

An example is shown below of a cutting list for a simple box (Fig. 1). The list shows the parts of the box, the kind of wood, the number required of each part and the finished size of the part. In the last column we find the size of the timber that will be required when the planing allowance has been added.

Usually it is best to plan a project according to the sizes of timber you have available, subtracting the planing allowances first in order to prevent waste.

Cutting list:

part	wood	no.	finished size (cm)			timber size (cm)		
			L	W	T	L	W	T
A	Odum	1	10,6	8	2,2	11,8	8,6	2,5
B	Odum	1	10,6	8	2,2	11,8	8,6	2,5
C	Odum	1	25	8	2,2	26,2	8,6	2,5
D	Odum	1	25	8	2,2	26,2	8,6	2,5

When you are choosing timber for a piece, choose the best, straight boards for the long pieces. The crooked or defective boards can be used for the shorter pieces.

NOTES:

Fig. 1

Fig. 2

Fig. 3

Fig. 4

Fig. 5

Fig. 6

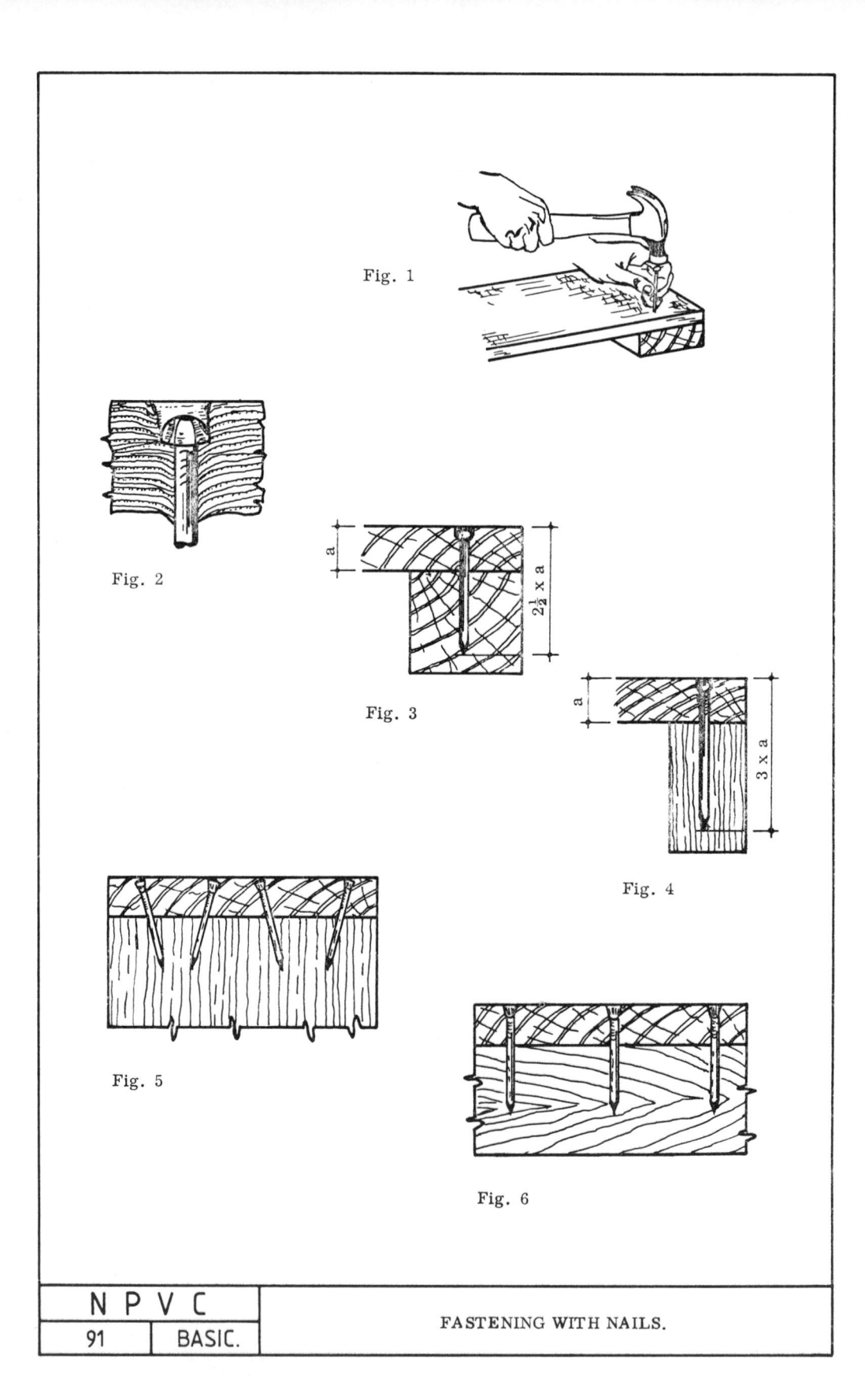

NPVC		FASTENING WITH NAILS.
91	BASIC.	

FASTENING WITH NAILS

Before you read this section, look up Nails in the Reference Book, Products section, page 207.

DRIVING NAILS

To start a nail, hold it steady between your thumb and fingers with one hand and strike one or two light blows with the hammer (Fig. 1).

After the nail is well started, drive it in with firm blows. Hold the handle of the hammer near the end and strike the nailhead straight.

When you drive a nail all the way in, be careful on the last blows not to hit the wood and leave a hammer mark on it.

HOLDING POWER

The holding power of the nail depends on the pressure of the wood fibres against the shank of the nail and also on the size of the nailhead (Fig. 2).

Hard dry wood holds better than soft or wet wood. End grain doesn't hold nails very well. If the nail is driven across the fibres, the nail's length should be $2\frac{1}{2}$ times the thickness of the top piece (Fig. 3). If it is driven into end grain, the length should be 3 times the thickness of the top piece (Fig. 4).

The holding power of nails in end grain can be improved by dovetail nailing, which means the nails are inserted at slight angles (Fig. 5) instead of straight in (Fig. 6) which is the usual way.

If possible the nails should be inserted at right angles to the force that will be applied to the piece; so that any force tends to shear off the nail rather than pull it out.

The correct placing of the nails is important with respect to the strength of the finished piece.

NOTES:

FASTENING WITH NAILS.

Fig. 1

Fig. 2

PUTTY

Fig. 3

Fig. 4

Fig. 5
a
SUPPORT
b
SUPPORT
c
SUPPORT

SUPPORT

Fig. 6

VICE

N P V C		FASTENING WITH NAILS.
93	BASIC.	

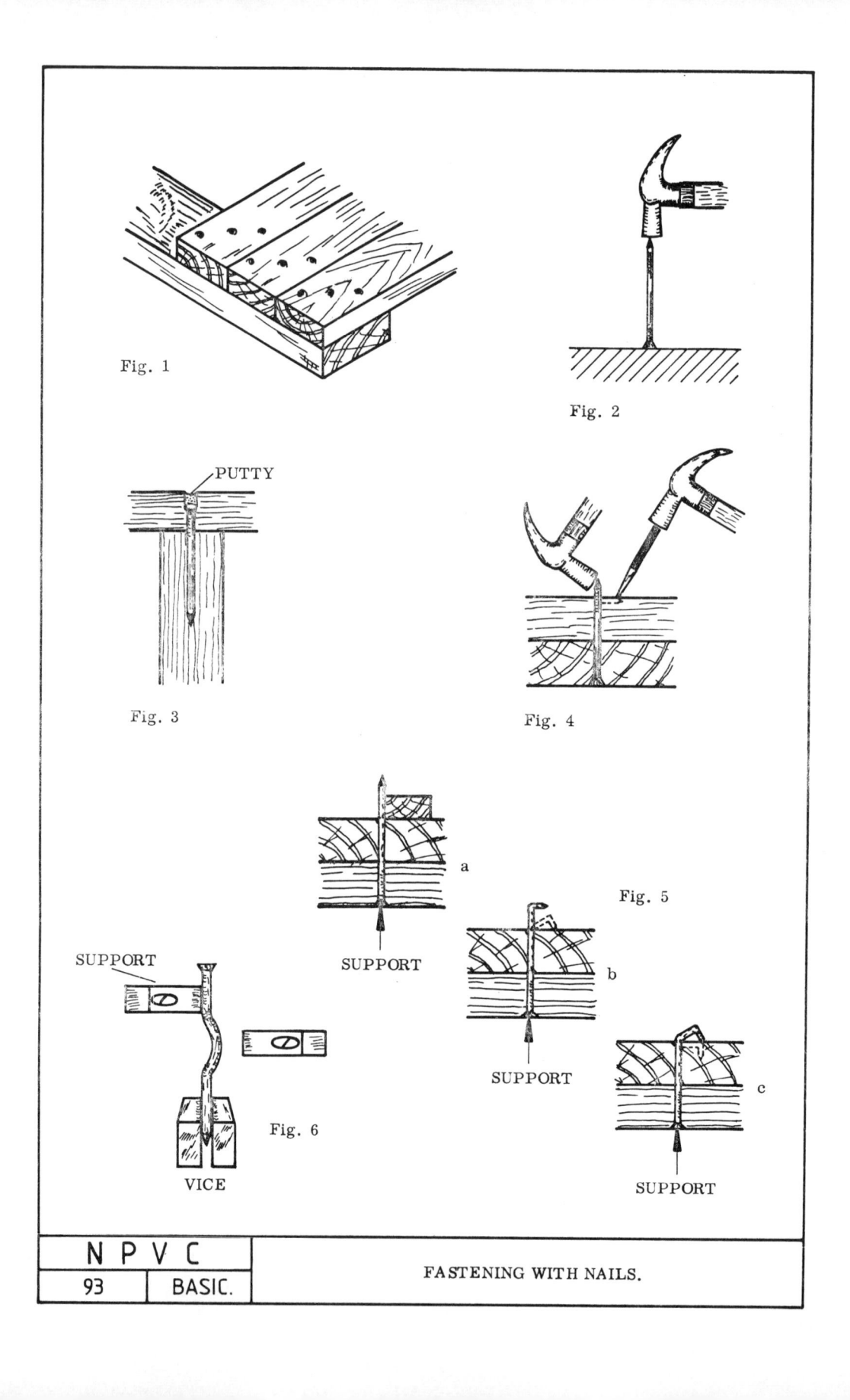

HOW TO PREVENT SPLITTING DURING NAILING

Nails can be staggered (inserted out of line) to prevent splitting along the fibres of the wood (Fig. 1).

Blunting the nail with one or two hammer blows on the tip also helps to prevent the nail from splitting the wood (Fig. 2).

If large nails are to be fixed, drill holes first to keep them from splitting the timber. The holes should be slightly smaller in diameter than the nails.

FINISHING OFF

Lost head nails are punched (knocked below the surface of the wood) with a nail punch or a large blunt nail. The remaining hole can be filled with putty (Fig. 3).

When the sharp points of nails come all the way through the timber and out on the other side, they are clenched; that is the tips are bent over and flattened against the wood, out of the way. There are two ways of doing this:

- Knock the tip flat and punch it into the wood with a nail punch (Fig. 4).

- Bend the point at a right angle first, and then knock it back into the wood (Fig. 5; a, b, & c). This is done where the nail projects more than 1 cm.

The head of the nail should be supported during clenching to keep it from being pushed out again.

Bent nails can be straightened if you tighten one end in a vice and support the other end with a hammer while knocking out the bent part with another hammer (Fig. 6).

NOTES:

Fig. 1

Fig. 2

Fig. 3

Fig. 4

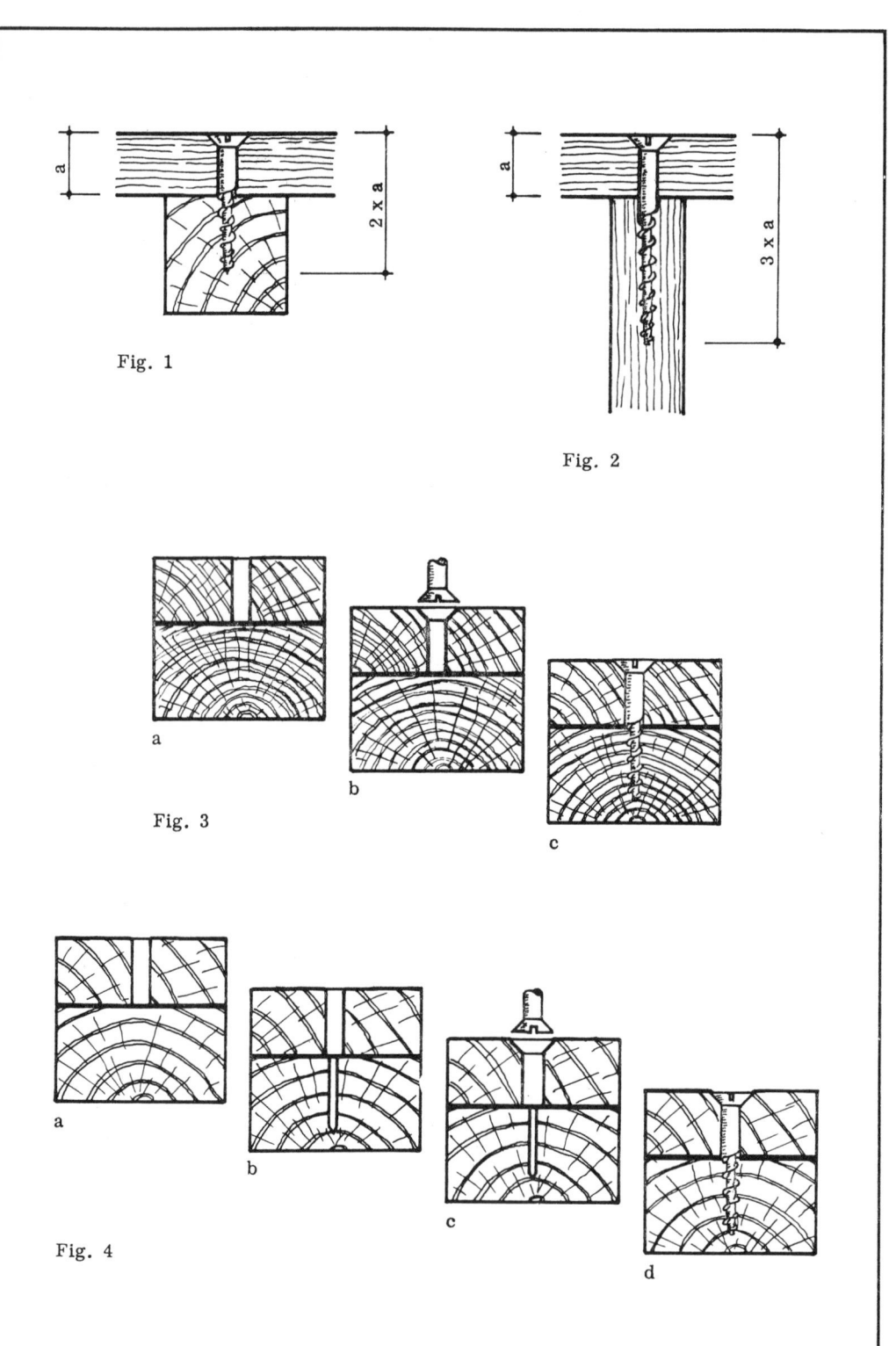

N P V C		FASTENING WITH SCREWS.
95	BASIC.	

FASTENING WITH SCREWS

Before you read this section, look up screws in the Reference Book, page 209; screwdrivers on page 77; and maintenance of screwdrivers on page 103.

HOLDING POWER

The holding power of a screw depends on how the thread embeds in the fibres, the length of the screw and the strength of the head which holds the top piece.

When a screw is driven across the grain, the screw's length should be about twice the thickness of the top piece (Fig. 1).

Screws driven into the end grain should be longer, about 3 times the thickness of the top piece (Fig. 2).

DRIVING SCREWS IN SOFT WOOD

- Bore a hole in the top piece with the same diameter as the shank of the screw (Fig. 3a). The bottom piece may be punched with a large nail or awl (Reference Book, page 73).
- Countersink if necessary (Fig. 3b).
- Drive the screw (Fig. 3c).

DRIVING SCREWS IN HARD WOOD

- Bore a hole in the top piece with the same diameter as the shank of the screw (Fig. 4a).
- Bore a hole in the bottom piece with the same diameter as the core of the screw (Fig. 4b).
- Countersink if necessary and drive the screw (Figs. 4c & 4d).

Use a screwdriver with the correct tip only. If the screw turns too hard, the hole may be too small or not deep enough. Remove the screw and find out what the problem is so it can be corrected, otherwise the screw may break or split the board.

- Turn the screw down until the head is just seated. Overturning weakens the holding power and may break the screw.
- To make driving easier and to protect against rust, a bit of soap or oil may be put on the tip of the screw.

FASTENING WITH SCREWS.

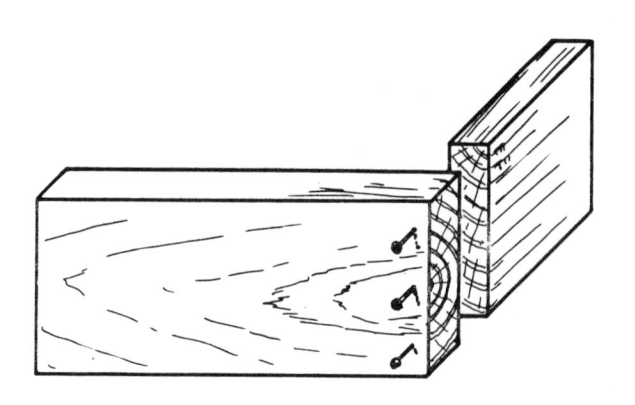

Fig. 1

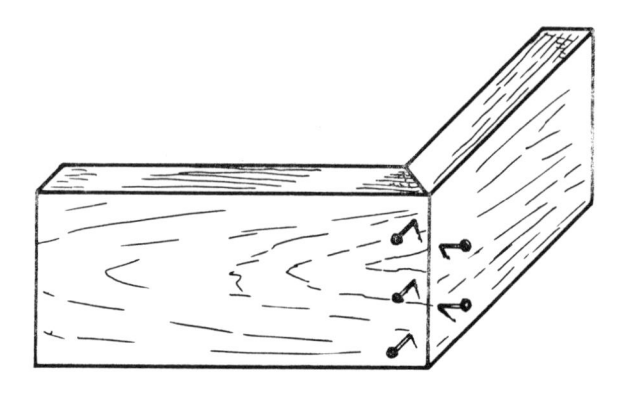

Fig. 2

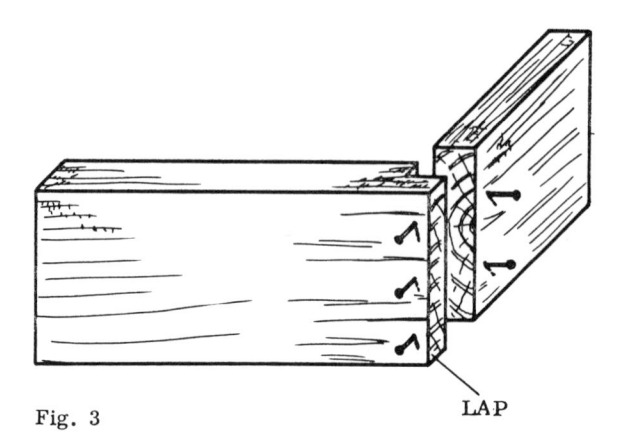

Fig. 3

LAP

N P V C		ANGLE JOINTS.
97	BASIC.	

ANGLE JOINTS

Angle joints are joints where the sides of the pieces (the wide surfaces) meet at right angles to each other. Angle joints are used for box-like constructions such as small boxes, tool boxes etc. For an example of a box using angle joints, see the illustration for the Cutting List lesson, on page 89. In this chapter we will consider the most common types of angle joints and their construction.

NAILED BUTT JOINT

The simplest angle joint is the nailed butt joint. The end of one piece of wood is cut square, then butted against the face of the other piece. It is held in place with nails, or both nails and glue (Fig. 1).

PLAIN MITRED JOINT

The ends of the pieces are mitred (cut at 45 degrees) across the thickness. The mitred ends are butted together and held in place with glue and nails (Fig. 2). This is a weak type of joint although it is stronger than the butt joint because it is nailed from two sides. Its advantage is that the end grain is not exposed to damage from water or insects, and it has a neater appearance.

REBATED BUTT JOINT

The end of one piece fits into a rebate at the end of the other piece. This joint is strong because two surfaces are available for nailing, and because the shoulder of the rebate supports and helps to hold the other piece (Fig. 3).

The lap is the section of wood which is left projecting after the rebate is cut (Fig. 3). The lap is usually one-half of the thickness of the board. This lap will be important later when we are figuring out the length of our pieces for making a box.

The rebated butt joint is simple to construct. In the following sequence of operations we will describe how to make a simple box with this joint while also practising some techniques discussed earlier, like making a cutting list and preparing timber to size.

The tools required here will be the same ones we used for the preparation of timber; with the additions of a firmer chisel, a smoothing plane, and a backsaw. Make sure you know what these tools are and how to use them before you go on.

Fig. 1

SIDE
END

REBATE POSITION

Fig. 2

SHOULDER LINE
DEPTH

Fig. 3

Fig. 4

Fig. 5

Fig. 6

PLANE FROM BOTH SIDES

Fig. 7

NPVC | 99 | BASIC. | ANGLE JOINTS.

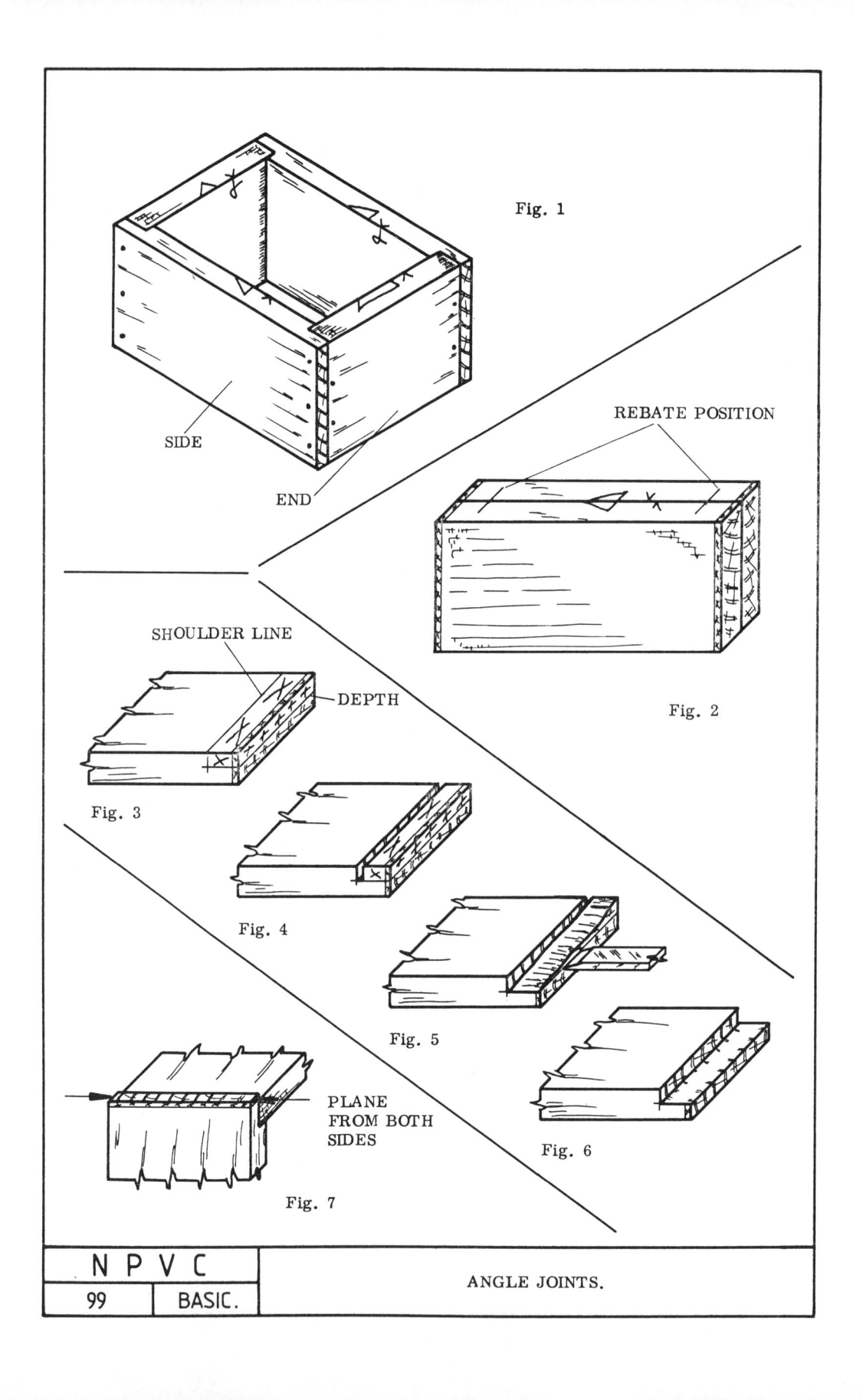

SEQUENCE OF OPERATIONS FOR CONSTRUCTING A BOX WITH THIS JOINT

Step 1. Preparation of timber

- Make a cutting list. The end pieces can be cut to the required length, that is the outside width of the box minus the width of the two laps. Allow 3 mm extra at each end of the side pieces, for planing off after assembly (Fig. 2).

- Prepare the pieces (see Preparation of Timber section, pages 84 to 86).

Step 2. Marking out

- Mark the sides and ends as shown in Fig. 1, on the face edges (Marking of frames, page 88). All further marking will be done from the sides with these marks.

- Place the two sides together and mark the position of the rebates, squaring with the try square (Fig. 2).

- Mark the shoulder lines of the rebate on the inside face of the piece, using the try square (Fig. 3).

- Mark the depth of the rebate on the end grain and the edge, using a marking gauge (Fig. 3). Show the waste with crosses.

Step 3. Cutting the rebate

- Saw the shoulders down to the gauge line. Cut on the waste side of the line. If the piece is very wide, nail or clamp a guide over the line to guide the saw. Use a backsaw (Fig. 4).

- Remove the waste carefully to the gauge line with a firmer chisel (Fig. 5). Find out the direction of the grain by chiselling out small pieces first, so that you don't accidentally chisel too deep.

Step 4. Assembling

- Clean up the inside faces with a smoothing plane (Fig. 6).

- Assemble the box with glue and nails.

- Measure the diagonals to check for squareness.

- Clean up the face and bottom edges with a smoothing plane.

- Plane off the waste from the sides with the smoothing plane. Prevent splintering by working inwards from the ends (Fig. 7).

NOTES:

ANGLE JOINTS.	N P V C	
	BASIC.	100

Fig. 1

Fig. 2

Fig. 3   DEPTH

Fig. 4

Fig. 5

NOTES:

N P V C		ANGLE JOINTS.
101	BASIC.	

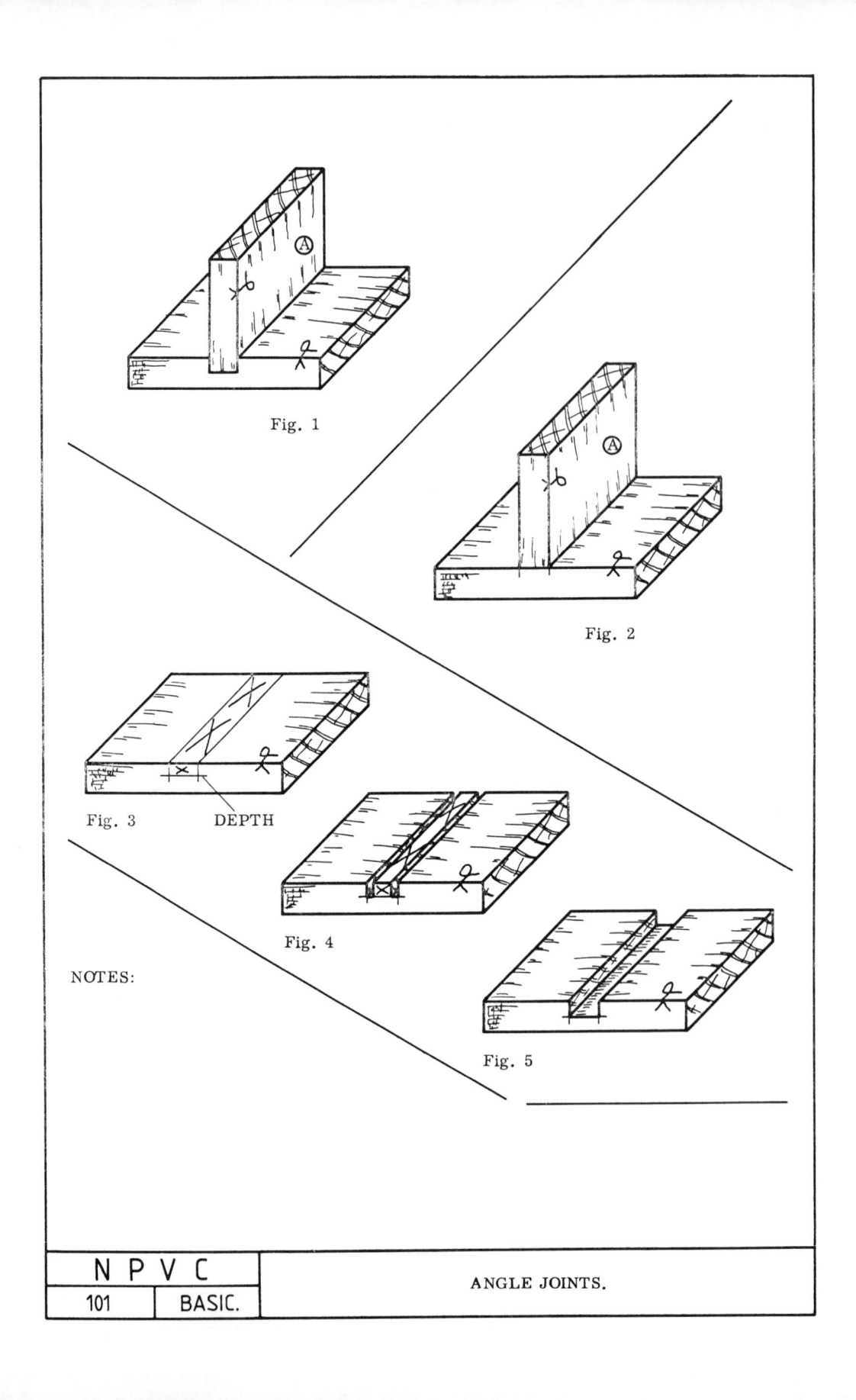

HOUSED JOINT

These joints are another type of angle joint, also used in box-like constructions.

Housing consists of sinking the end of one piece into a trench which is cut into the face of another piece (Fig. 1).

The tools required for making this type of joint are the same ones used to make the rebated butt joint.

SEQUENCE OF OPERATIONS FOR CONSTRUCTING THE JOINT

Step 1. Preparation of the timber

- Make a cutting list.

- Prepare the timber (see Preparation of Timber section).

Step 2. Marking out

- Mark one edge of the trench with a try square and the other edge by using piece A as a guide (Fig. 2). (Smooth piece A before using it to mark the trench).

- Gauge the depth of the trench at each edge (Fig. 3).

- Show the waste with small crosses (Fig. 3).

Step 3. Cutting the trench

- Saw the sides of the trench (on the waste side of the lines) down to the gauge lines (Fig. 4).

- Chisel out the waste from the trench (Fig. 5).

Step 5. Assembling

- Assemble the two parts with nails and glue.

- Clean up the edges with a smoothing plane.

NOTES:

ANGLE JOINTS.

MORTICE

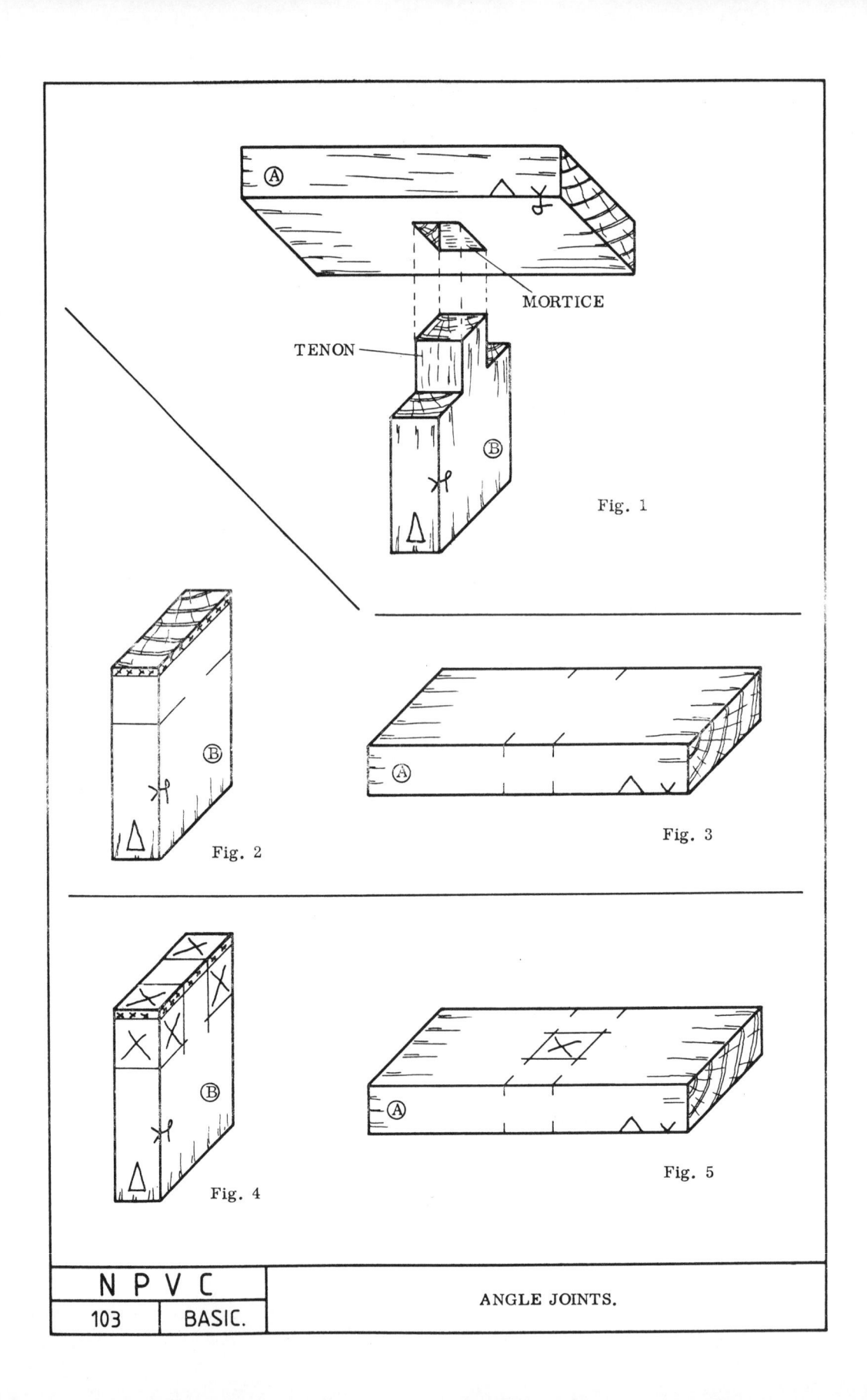

TENON

Fig. 1

Fig. 2

Fig. 3

Fig. 4

Fig. 5

N P V C	
103	BASIC.

ANGLE JOINTS.

COMMON MORTICE AND TENON JOINT FOR BOX-LIKE CONSTRUCTIONS

This is one of the commonest and strongest joints. The two parts are (Fig. 1): the tenon (B) which is a projection on the end of one part and the mortice (A), the hole in the other part into which the tenon fits. The tenon is usually 1/3rd of the width of the board.

The tools we need are ones we have discussed before, plus a mortice chisel, a brace and drilling bits.

SEQUENCE OF OPERATIONS FOR CONSTRUCTING THE JOINT

Step 1. Preparation of timber

- Make a cutting list.
- Prepare the timber to the required sizes. (In the following steps, the piece with the mortice is "piece A" and the one with the tenon is "piece B".)

Step 2. Marking out

- Mark out the length of the tenon on piece B. Allow 3 mm waste in the length and make square lines all around with a try square and pencil (Fig. 2).
- Take piece A and mark out the position of the mortice on the face edge and make square lines on the edges on both sides with the try square (Fig. 3).
- Set the marking gauge to the width of the tenon and mark the lines around piece B at the width. Mark the waste with small crosses (Fig. 4).
- Use the same setting to mark both faces of piece A and use a try square and (already smoothed) piece B to mark the remaining two lines for the width of the mortice (Fig. 5). Mark the waste with a small cross.

If the marking gauge has two pins, set each at its correct measurement and mark both lines at once. If not, mark with the first setting on all the members, then change the setting and mark the other measurement on all the members.

- Always mark from the face edge. Check the marking by setting piece B against the marks on piece A to see if they fit. Piece B must be smoothed first.

NOTES:

ANGLE JOINTS.

N P V C

BASIC. 104

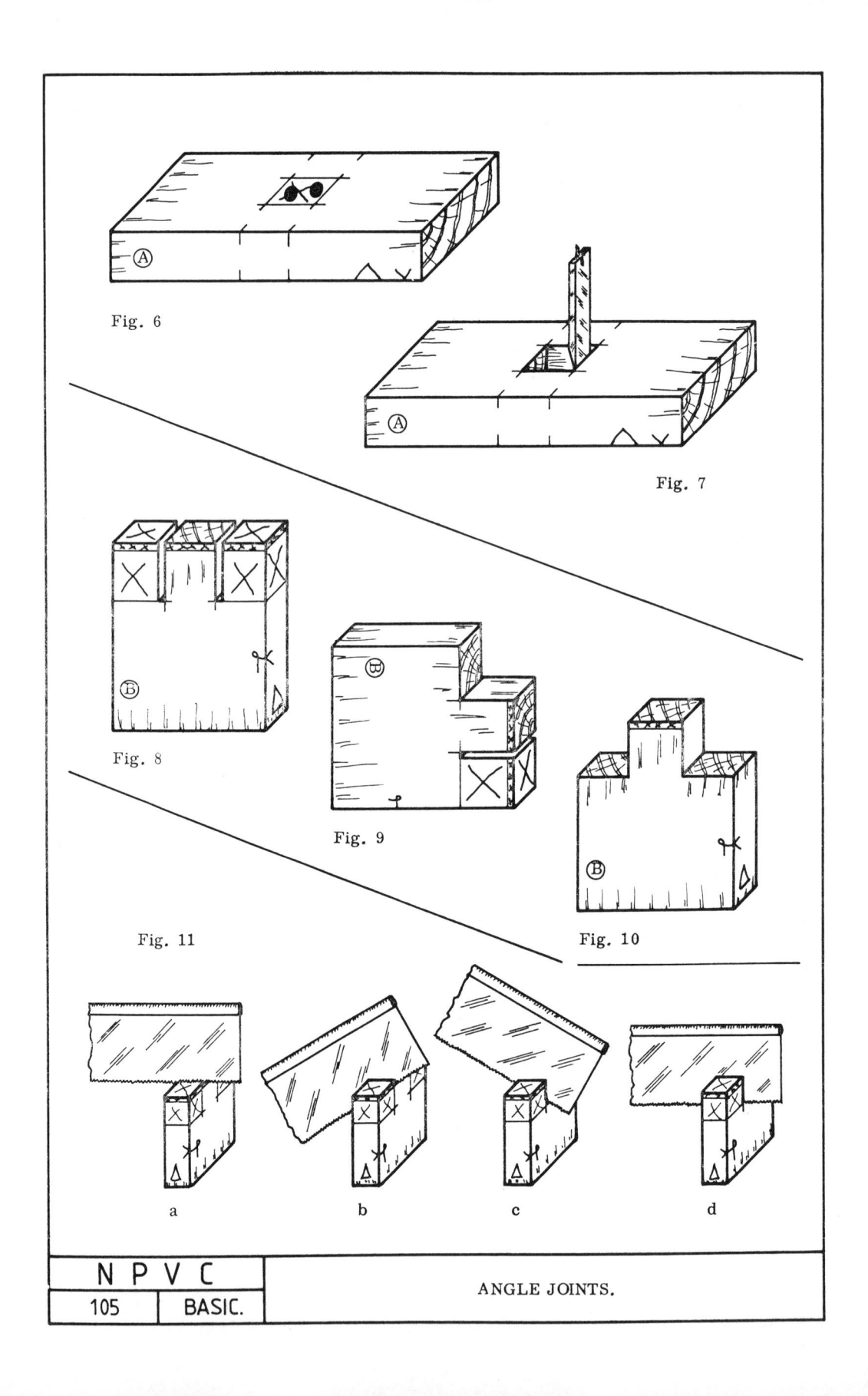

Fig. 6

Fig. 7

Fig. 8

Fig. 9

Fig. 10

Fig. 11

a   b   c   d

NPVC
105 BASIC.

ANGLE JOINTS.

Step 3. Cutting the mortice

- Bore out most of the waste, using a brace and bit (Fig. 6). Clamp a piece of wood to the underside to prevent splintering and damage to the bench.

- Chop out the remaining waste with a mortice chisel, chiselling halfway through from both sides. Leave about 2 mm extra waste on all sides to prevent damage to the sides. Keep the cutting edge of the chisel across the grain.

- Carefully chop out the rest of the mortice up to the lines (Fig. 7). Keep the bevel of the chisel towards the inside of the mortice. Do not use the mallet.

Step 4. Cutting the tenon

- Rip the sides of the tenon, sawing on the waste side of the line (Fig. 8). Cut in stages as shown in Fig. 11, a, b, c, & d).

- Carefully saw the shoulders, making sure to hold the saw straight. Keep on the waste side of the line (Figs. 9 & 10).

Step 5. Assembling the joint

- Check the fit of the members. The tenon should fit tightly into the mortice without splitting the morticed piece. There should be no gap between the shoulders of the tenon and the morticed member. Don't force the members together. If they don't fit, find the problem and correct it.

- Clean up the inside of the joint where it can't be reached after assembly with a smoothing plane. (Remember that the tenon should be smoothed before using it to mark out.)

- Assemble the joint.

- Plane off the waste end of the tenon, clean up all sides and edges with the smoothing plane.

NOTES:

ANGLE JOINTS.

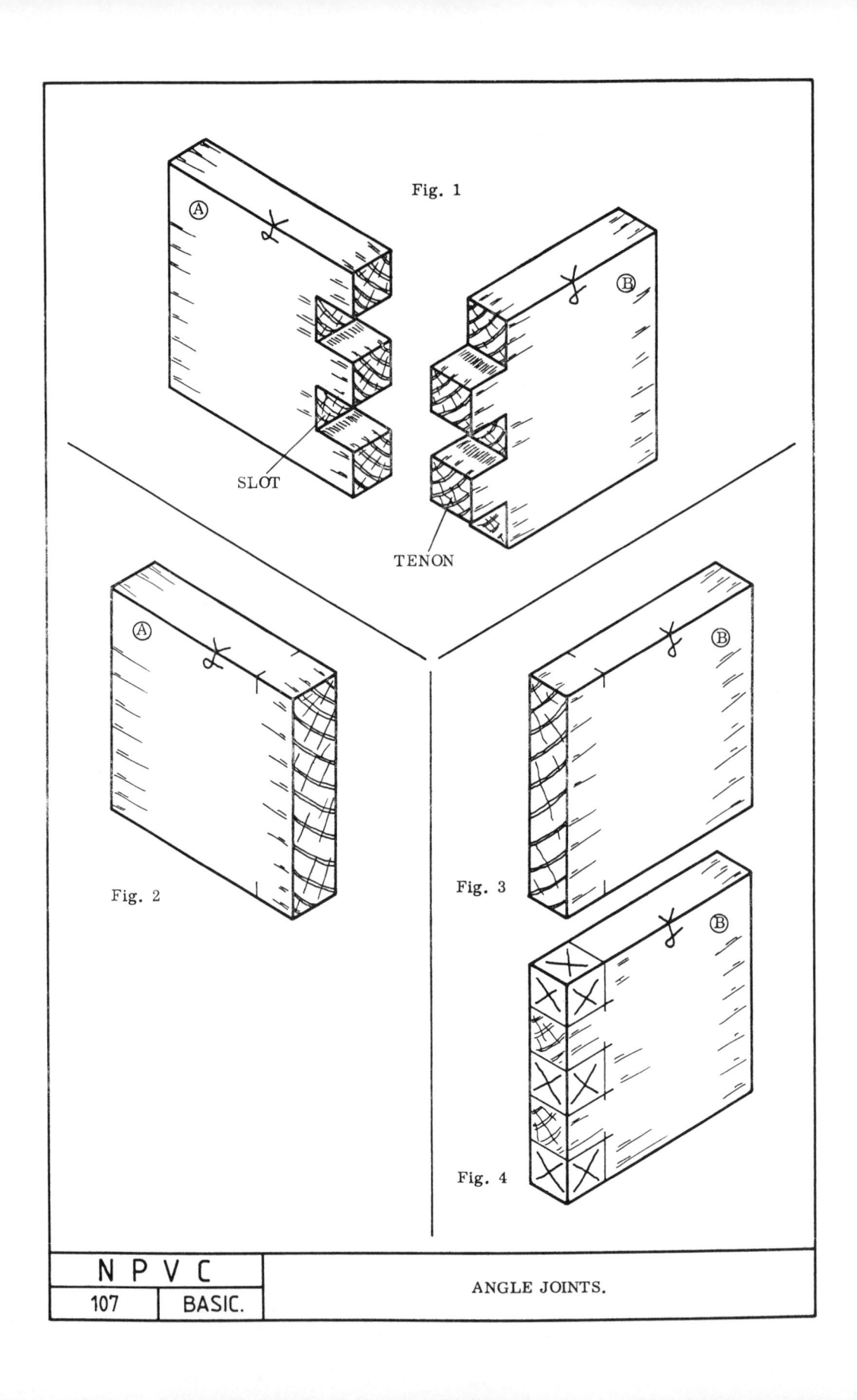

Fig. 1

Ⓐ

SLOT

Ⓑ

TENON

Fig. 2

Ⓐ

Fig. 3

Ⓑ

Fig. 4

Ⓑ

N P V C	
107	BASIC.

ANGLE JOINTS.

CORNERLOCKED JOINT

The cornerlocked joint is similar to the mortice and tenon joint. It is an angle joint with a series of tenons on one member which correspond to slots on the other member (Fig. 1). The resulting joint is strong because it can be nailed from two sides, and the interlocking tenons and slots also help hold the pieces together.

The tools required to make this joint are the same ones used for the mortice and tenon joint.

SEQUENCE OF OPERATIONS FOR CONSTRUCTING THE JOINT

Step 1. Preparation of the timber

- Make a cutting list.
- Prepare the timber to the required sizes. (In the following steps, the member with the slots is piece "A" and the one which has the tenons is piece "B".)
- If the members are to be used for a box where the external appearance is important, the face sides should be outside.
- In most cases the face edges are kept upwards.

Step 2. Marking out

- Mark out the position of the tenons and slots by gauging or squaring lines at the corners on the ends of the pieces: on piece A the depth should be equal to the thickness of piece B (Fig. 2); while on piece B the depth should be equal to the thickness of piece A (Fig. 3). Allow 2 mm waste for cleaning up after assembly.
- Mark out the shape of the tenons on piece B. Keep all tenons the same size (Fig. 4).
- Immediately mark the waste between the tenons with crosses (Fig. 4).

NOTES:

ANGLE JOINTS.

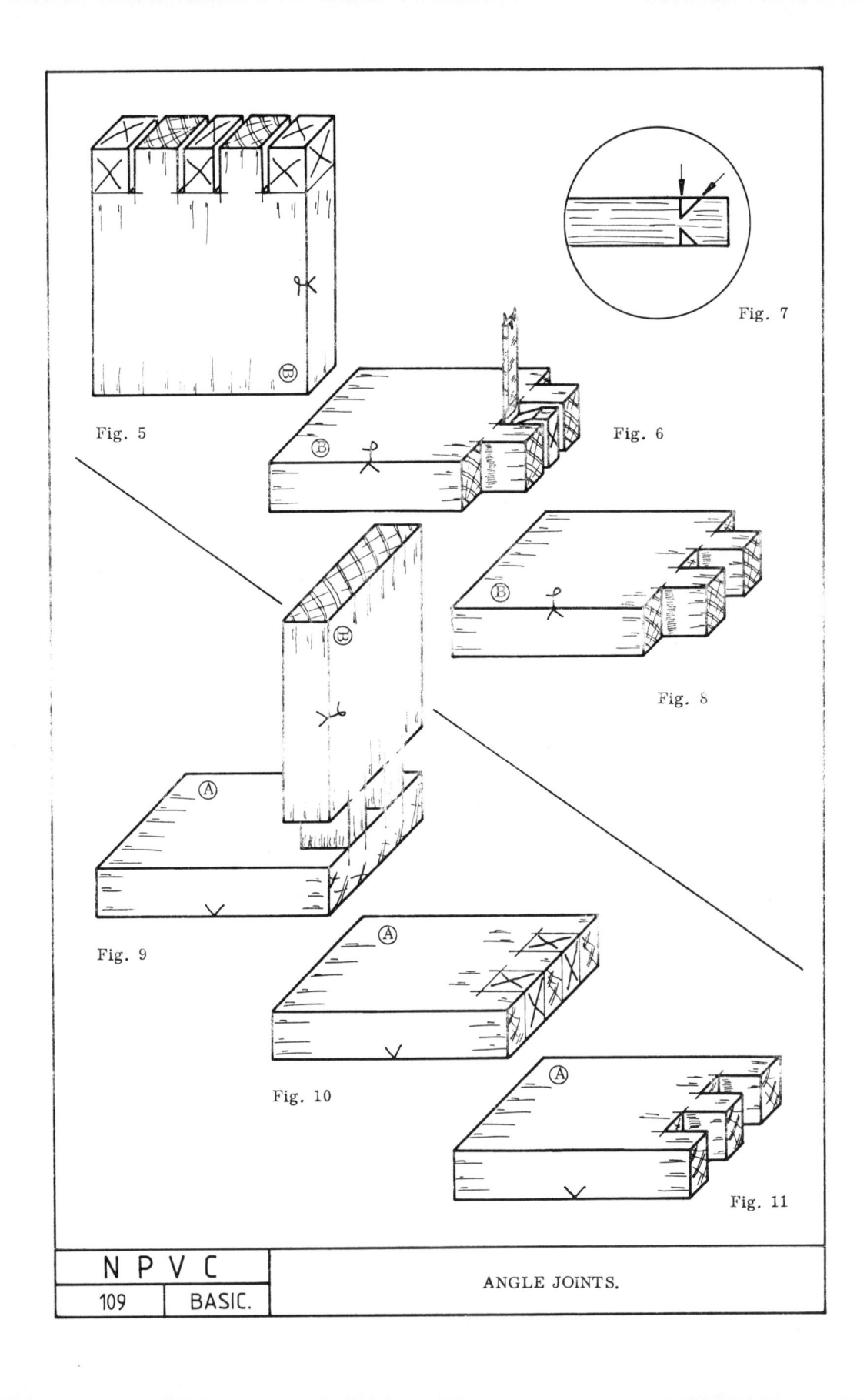

Fig. 5

Fig. 6

Fig. 7

Fig. 8

Fig. 9

Fig. 10

Fig. 11

N P V C	
109	BASIC.

ANGLE JOINTS.

Step 3. Cutting the tenons

- Rip the sides of the tenons down to the gauge line (Fig. 5). Saw on the waste side of the line.

- Chop out the waste by chiselling alternately vertically and then at an angle, making "V" cuts halfway through from each side (Figs. 6, 7, & 8).

Step 4. Cutting the slots

- Place piece B (with the tenons) over the end of piece A, with the face side towards the outside as indicated in Fig. 9.

- Mark the shape of the tenons onto piece A with a pencil (Fig. 9).

- Square the sides of the slots down both sides (Fig. 10).

- Mark the waste with small crosses (Fig. 10).

- Rip the sides of the slots, sawing on the waste side of the line.

- Chop out the waste from the slots, chiselling from both sides as explained in the previous step (Fig. 11).

Step 5. Assembling the joint

- Clean up the inside faces of the joint.

- Assemble the joint with glue and nails.

- When the glue is dry, clean up the waste of the tenons and slots with a smoothing plane. Make sure the nails are punched well below the surface to prevent damage to the sole of the plane.

- Clean up the outside faces and edges with a smoothing plane.

NOTES:

ANGLE JOINTS.

Fig. 1

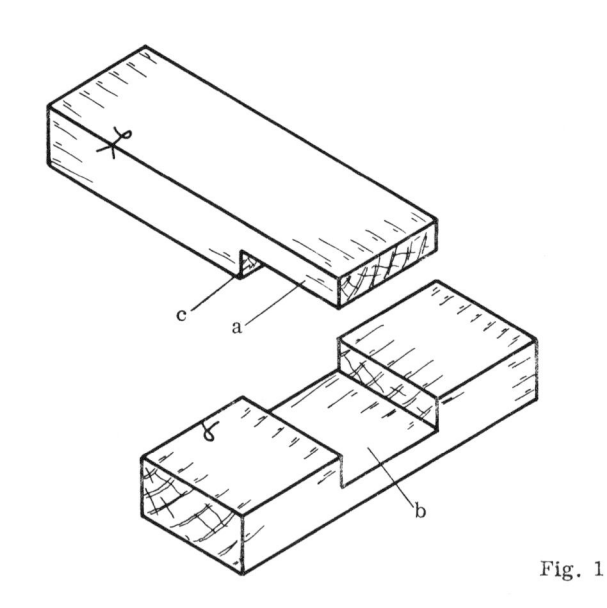

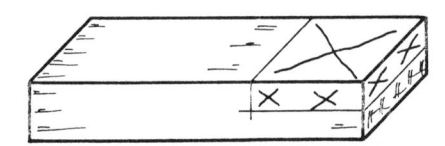

Fig. 2

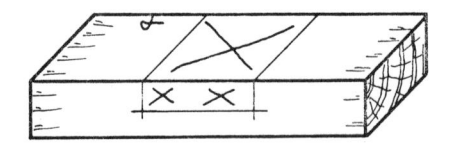

Fig. 3

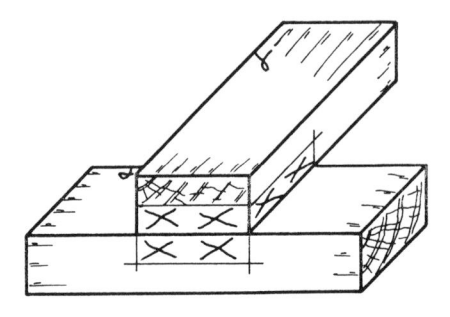

Fig. 4

N P V C		FRAMING JOINTS.
111	BASIC.	

FRAMING JOINTS

Framing joints are those used in frame-like constructions. The members are usually constructed with their edges at right angles to each other; in contrast to the angle joints used in box-like constructions, where it is the sides which form the right angle (previous pages).

HALVED JOINTS

Halved joints are one type of framing joint. The name is applied to joints where the pieces of timber which meet or cross each other are halved; that is, at the place where they cross, each piece is 1/2 the thickness of the rest of the piece. The result is that in the assembled joint, the surfaces of both pieces are flush.

Halved joints are used for constructing simple frames.

In Rural Building, we deal with four different kinds of halved joints. Here we will cover the description and construction of the "tee-halved joint", since the construction of the other joints follows much the same procedures.

The tee-halved joint consists of a pin (a) on the end of one piece which fits into a socket (b) in the other piece (Fig. 1).

The pin is half the thickness of the timber, and the depth of the socket equals the thickness of the pin. The shoulder of the pin (c) fits against the face edge of the socket (Fig. 1).

The tools required to make this joint are the same ones we used to make the mortice and tenon joint in the last chapter.

SEQUENCE OF OPERATIONS FOR CONSTRUCTING THE JOINT

Step 1. Preparation of timber

- Make a cutting list.

- Prepare the pieces to the required size.

Step 2. Marking out

- Mark the length of the pin by placing the socket piece on top of it and marking at the width. A small amount of waste can be left on the end of the pin, to be planed off after the joint is assembled.

- Make lines square at the shoulder of the pin, drawing them across the side and halfway down the edges, with a try square and pencil (Fig. 2). Mark the waste.

Fig. 5

a
b
c
d

Fig. 6

Fig. 7

Fig. 8

Fig. 9

N P V C	FRAMING JOINTS.
113	BASIC.

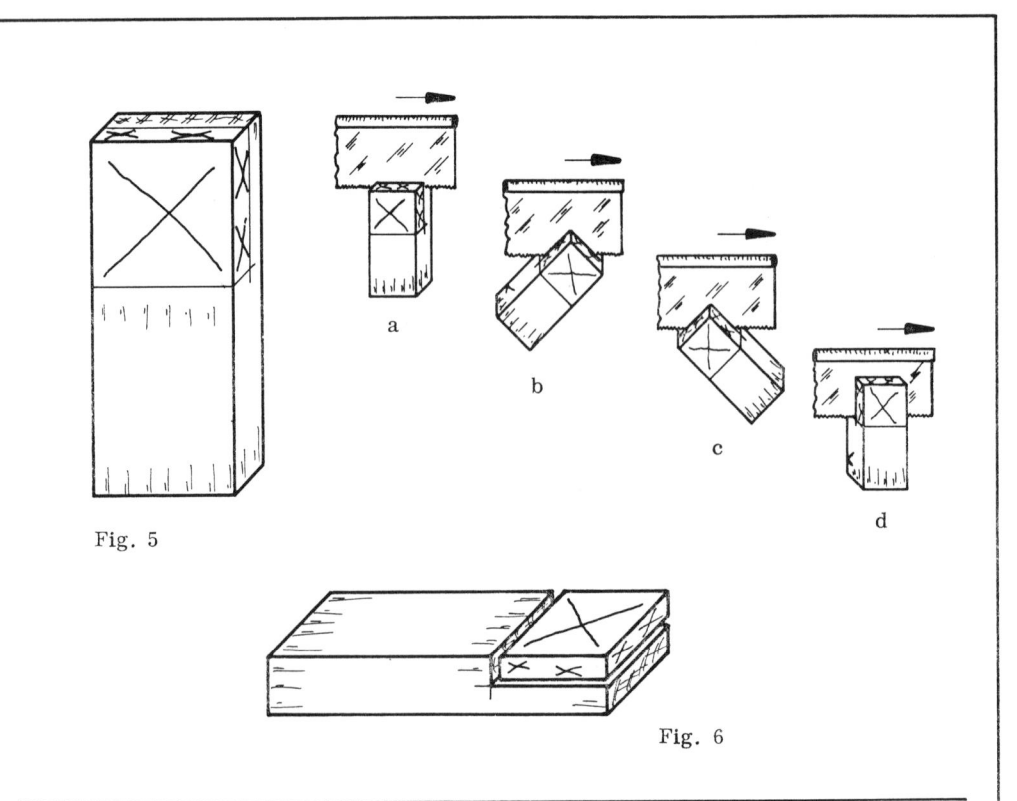

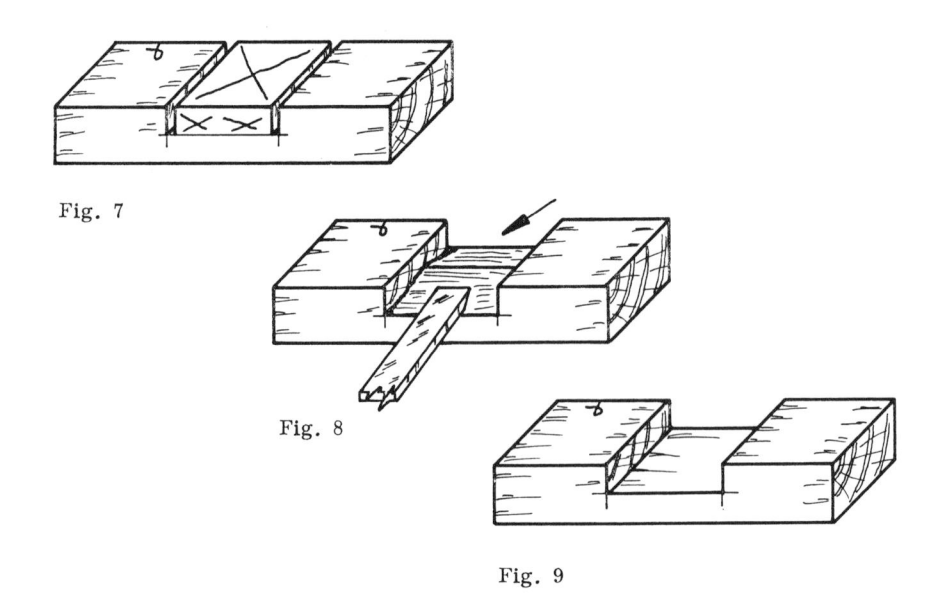

- Mark the position of the socket, using the piece with the pin as a guide. Smooth the pin before using it to mark the socket.
- Square the lines across the side and halfway down the edges with a try square. Mark the waste (Fig. 3, previous page).
- Gauge the thickness of the pin around its edges and mark the waste (Fig. 2, previous page).
- With the same setting, gauge the depth of the socket on both edges and mark the waste (Fig. 3, previous page).
- Both pin and socket should be gauged from the face side.
- Place the pin over the position of the socket and check the fitting (Fig. 4, previous page).

Step 3. Cutting the pin

- Rip the thickness of the pin. Cut in stages as shown in Fig. 5, a through d. Take care to keep on the waste side of the line.
- Saw the shoulder of the pin, keeping on the waste side of the line (Fig. 6).

Step 4. Cutting the socket

- Saw down to the gauge lines of the socket, keeping on the waste side of the lines (Fig. 7).
- Chisel out the waste, chiselling halfway through from both edges (Figs. 8 & 9).
- Test the flatness of the socket with the blade of the try square.

Step 5. Assembling the joint

- Clean up the inside edges with a smoothing plane.
- Assemble the joint with glue and nails.
- When the joint is dry, plane off the waste of the pin.
- Clean up all sides and edges with the smoothing plane.

NOTES:

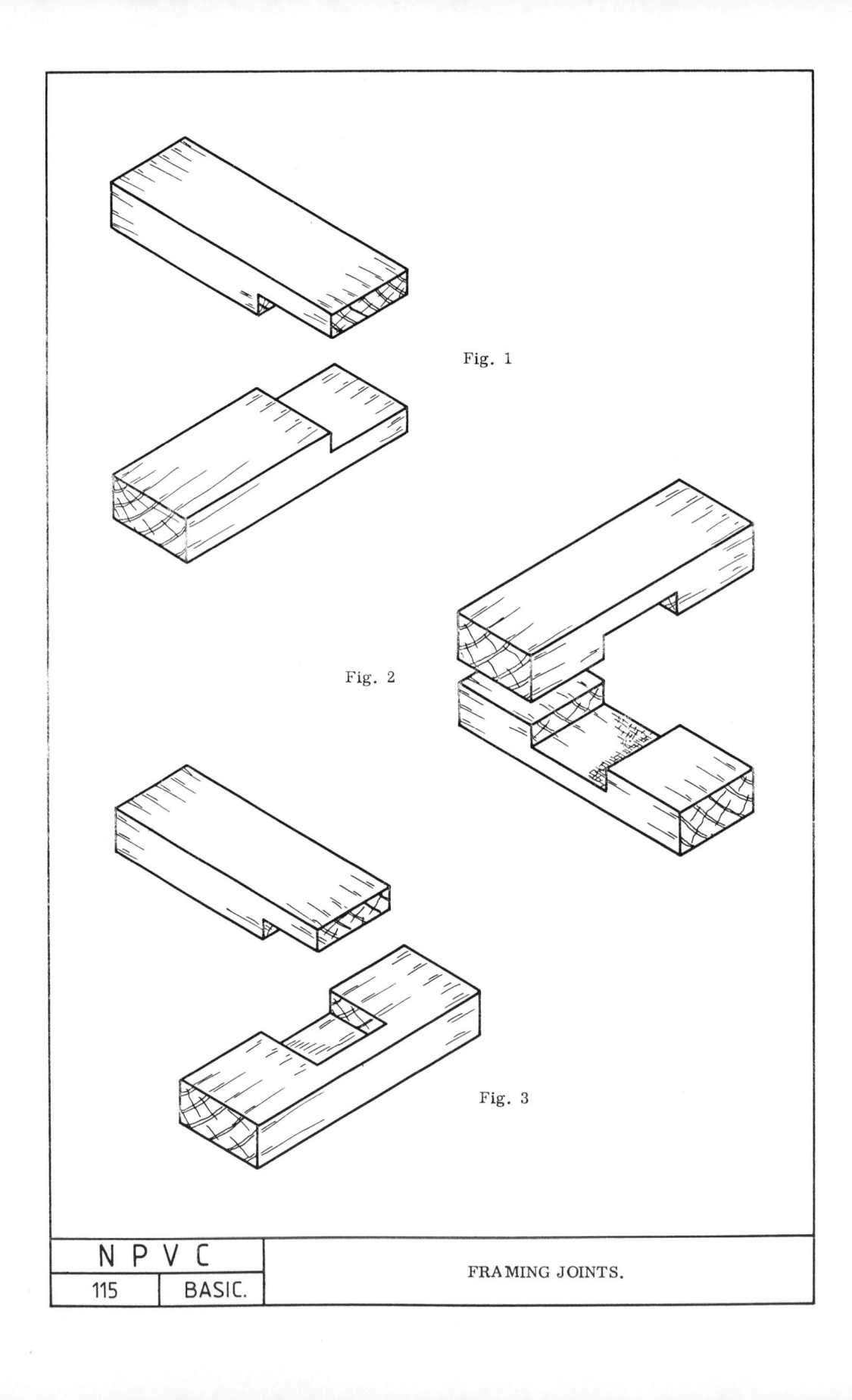

Fig. 1

Fig. 2

Fig. 3

NPVC		FRAMING JOINTS.
115	BASIC.	

CORNER-HALVED JOINT

Another halved joint is the corner-halved joint (Fig. 1). It is used where the pieces meet at their ends to form a corner.

The sequence of operations to construct this joint is similar to the one for the tee-halved joint, except that instead of a pin and a socket, two pins have to be marked and cut.

CROSS-HALVED JOINT

The third halved joint we deal with is the cross-halved joint (Fig. 2). It is used where two members cross each other.

The sequence of operations to construct this joint is similar to the tee-halved joint, but instead of a pin and a socket, two sockets have to be marked and cut.

STOPPED TEE-HALVED JOINT

In this joint the socket is stopped away from the edge and the pin is cut short, so that in the assembled joint the end grain of the piece is not seen (Fig. 3).

Otherwise, the same sequence is followed as for the tee-halved joint.

NOTES:

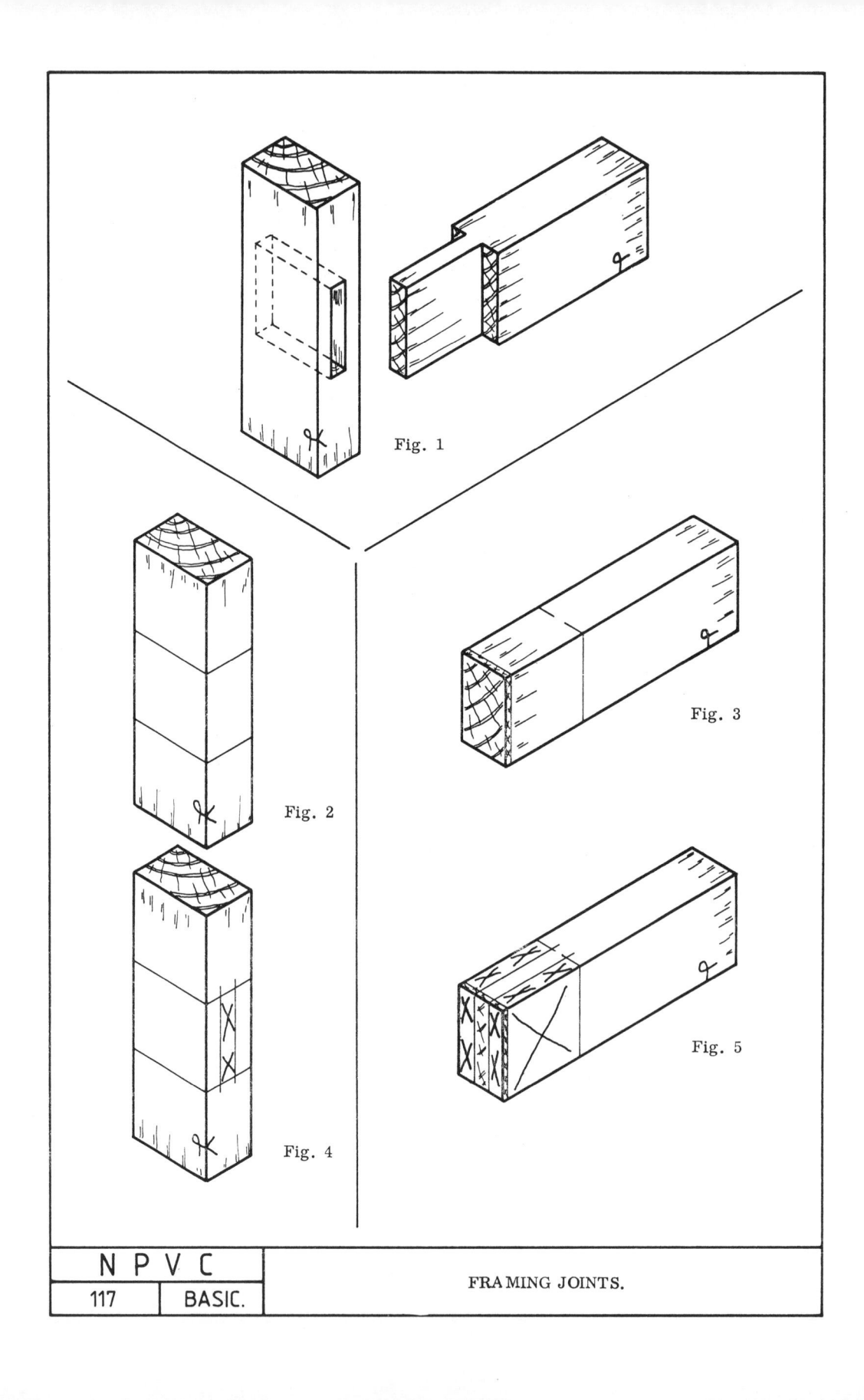

Fig. 1

Fig. 2

Fig. 3

Fig. 4

Fig. 5

N P V C		FRAMING JOINTS.
117	BASIC.	

COMMON MORTICE AND TENON JOINT FOR FRAME-LIKE CONSTRUCTIONS

One of the most common and strongest forms of framing joint is the mortice and tenon joint (Fig. 1).

The sequence of operations to construct a mortice and tenon joint for frame-like constructions is almost the same as for box-like constructions. Of the four types of mortice and tenon joints mentioned in this chapter, we will only go into detail about the construction of one of them, the common mortice and tenon. No new tools will be needed.

SEQUENCE OF OPERATIONS FOR CONSTRUCTING THE JOINT

Step 1. Preparation of timber

- Make a cutting list.
- Prepare the timber.

Step 2. Marking out

- Mark out the position of the mortice and square the lines across the face side and edges, using a try square and pencil (Fig. 2).
- Mark out the length of the tenon on the other member. Allow 3 mm waste on the end.
- Square lines all around (Fig. 3).
- Set a marking gauge to the size of the tenon (one-third of the width of the piece) and mark around the end of the tenon (Fig. 5). Mark the waste.
- Use the same setting to mark both edges of the mortice and mark the waste (Fig. 4).
- Do all marking from the face side.
- Check the marking, using the pieces as a guide by placing them over the marks (compare this sequence to the mortice and tenon for box-like constructions, page 104).

NOTES:

FRAMING JOINTS.

Fig. 6

Fig. 7

Fig. 8

Fig. 9

Fig. 10

Fig. 11

NPVC		FRAMING JOINTS.
119	BASIC.	

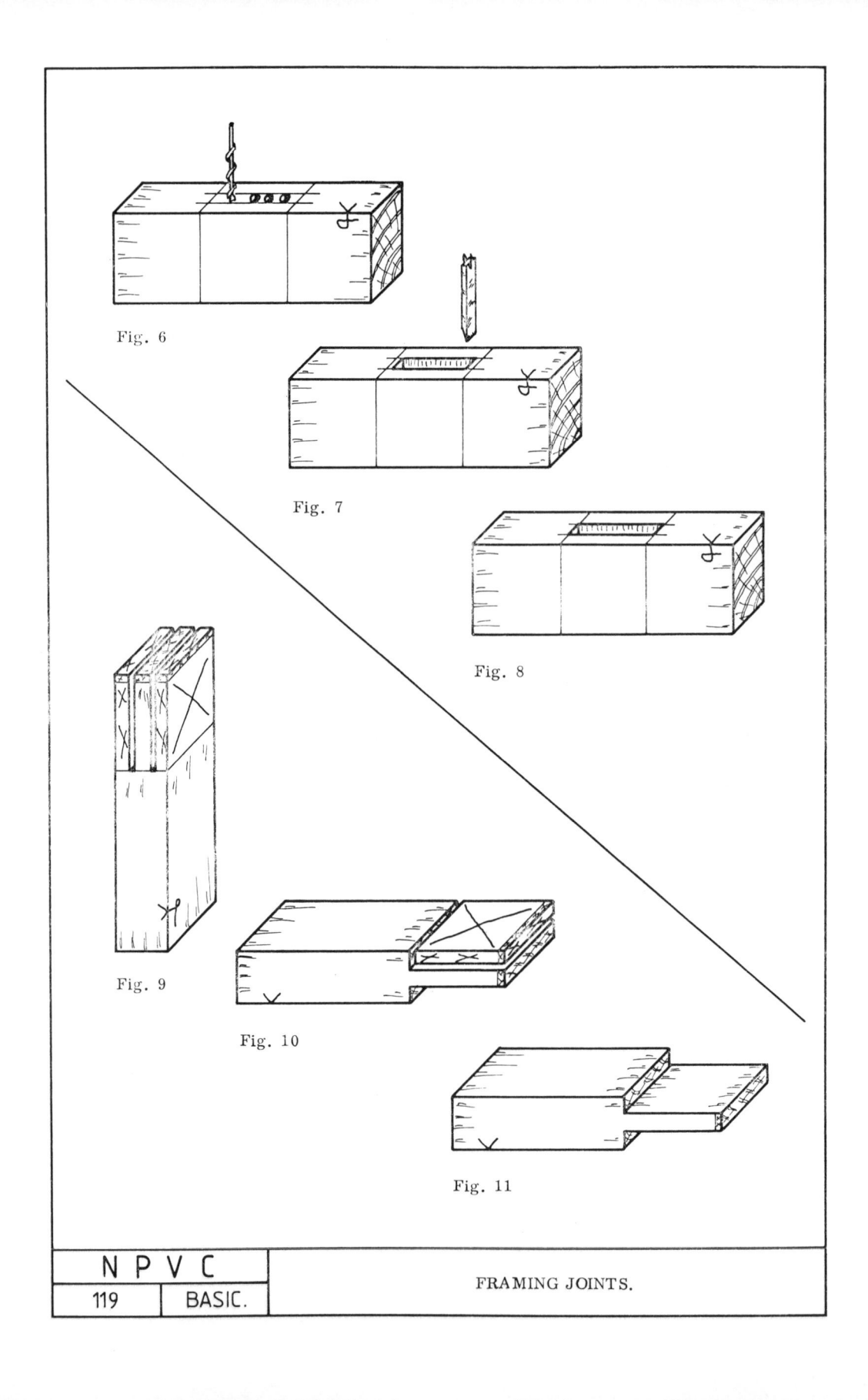

Step 3. Cutting the mortice

- Most of the waste may be bored out (Fig. 6). Bore halfway through from both edges. Make sure you keep the brace at a 90 degree angle to the edge.
- Chop out the remaining waste, chiselling halfway through from both edges. Leave about 2 mm extra to prevent damage to the sides of the mortice during chiselling (Fig. 7).
- When most of the waste is out, chisel out the remainder to the line (Fig. 8).
- Keep the cutting edge of the chisel across the grain.

Step 4. Cutting the tenon

- Rip the sides of the tenon, sawing on the waste side of the lines (Fig. 9).
- Saw in steps (see tee-halved joint).
- Carefully saw the shoulders, keeping the saw vertical and on the waste side of the line (Fig. 10 & 11).

Step 5. Assembling the joint

- Check whether the members fit together (see Assembly section for the mortice and tenon joint for box-like constructions).
- Clean up inside the joint where it cannot be reached after assembly.
- Assemble the joint with glue.
- When it is dry, plane off the waste of the tenon.
- Clean up the edges and sides with a smoothing plane.

Note the importance of marking the waste as you mark out the pieces. This cannot be over-emphasized. Most construction mistakes are made by cutting on the wrong side of the line, due to improper marking.

NOTES:

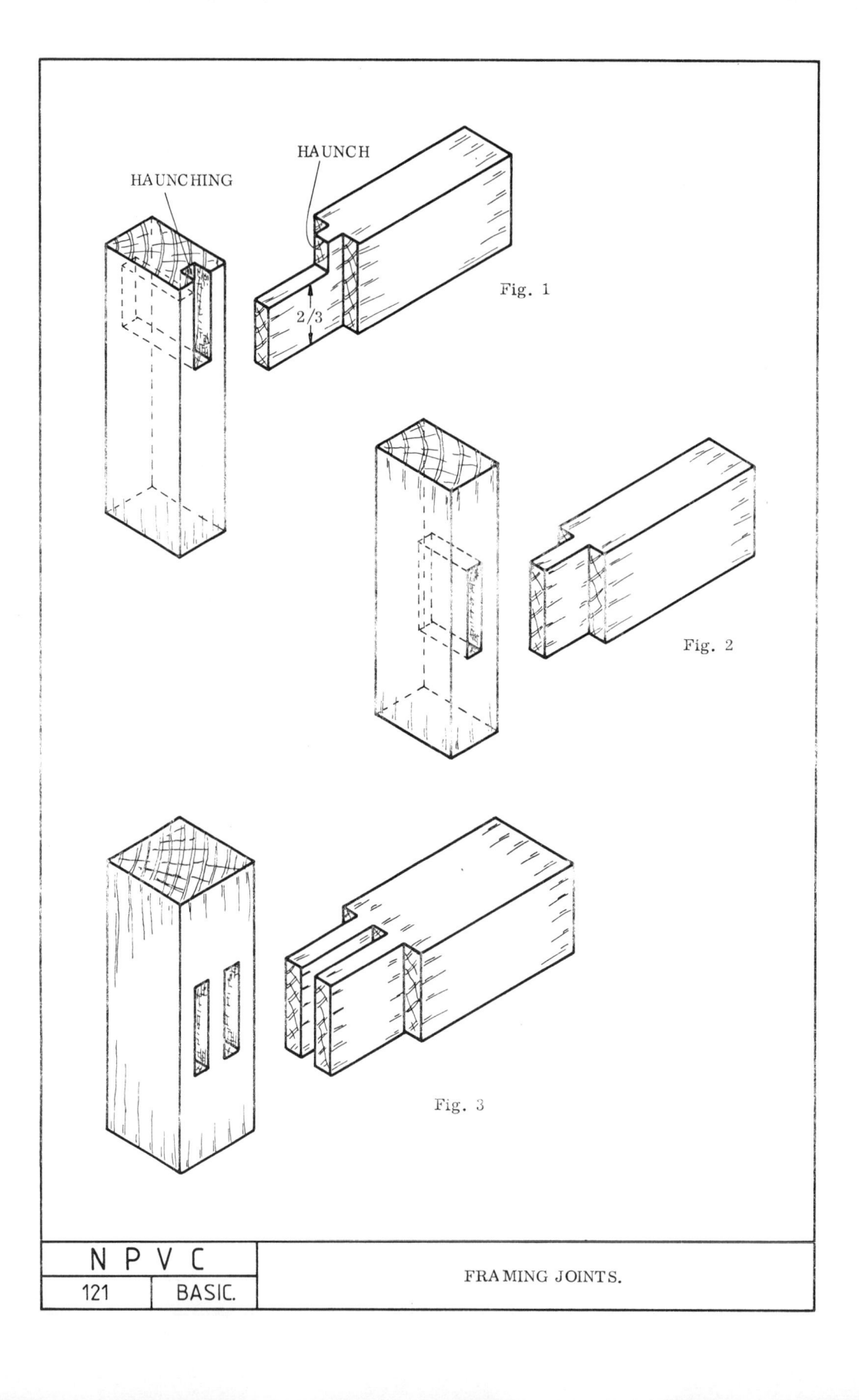

HAUNCHING

HAUNCH

2/3

Fig. 1

Fig. 2

Fig. 3

N P V C		FRAMING JOINTS.
121	BASIC.	

HAUNCHED MORTICE AND TENON JOINT

Another type of mortice and tenon for frame-like constructions is the haunched mortice and tenon (Fig. 1). This joint is used where one member meets another at a corner.

The width of the tenon is reduced to 2/3rd of the width of the board and the mortice size is reduced to suit (Fig. 1).

A haunch is left on the tenon to prevent it from twisting in the mortice. The length of the haunch is equal to the thickness of the tenon and it fits into a recess above the mortice, called the haunching.

Otherwise, the sequence of operations for construction of this kind of joint is the same as for the common mortice and tenon joint.

When you make the cutting list for this type of joint, the allowance in length for the member with the mortice should be 25 mm instead of 12 mm to help prevent splitting of the haunching (see Cutting List, page 90).

STUB TENON JOINT

Where the end grain of the tenon and the opening of the mortice must be hidden, the stub tenon joint is chosen (Fig. 2). In this joint the tenon does not pass through the morticed member, but is stopped inside. The sequence of operations for constructing this joint is the same as for the common mortice and tenon joint. Stub tenons are also used for box-like constructions.

At times a combination of the haunched and stub tenons is required. This is called a haunched stub mortice and tenon joint.

TWIN TENON JOINT

Where the members to be joined are very thick, twin tenons are used (Fig. 3). Each tenon is then not 1/3rd, but 1/5th of the thickness of the members.

The sequence of operations is almost the same as for the common mortice and tenon joint, with the only difference being that two mortices and tenons have to be marked and cut instead of only one.

This joint can be used for both frame-like and box-like constructions.

NOTES:

Fig. 1

a
b
c
d
e

WEDGES

HOLES TO PREVENT SPLITTING

Fig. 2

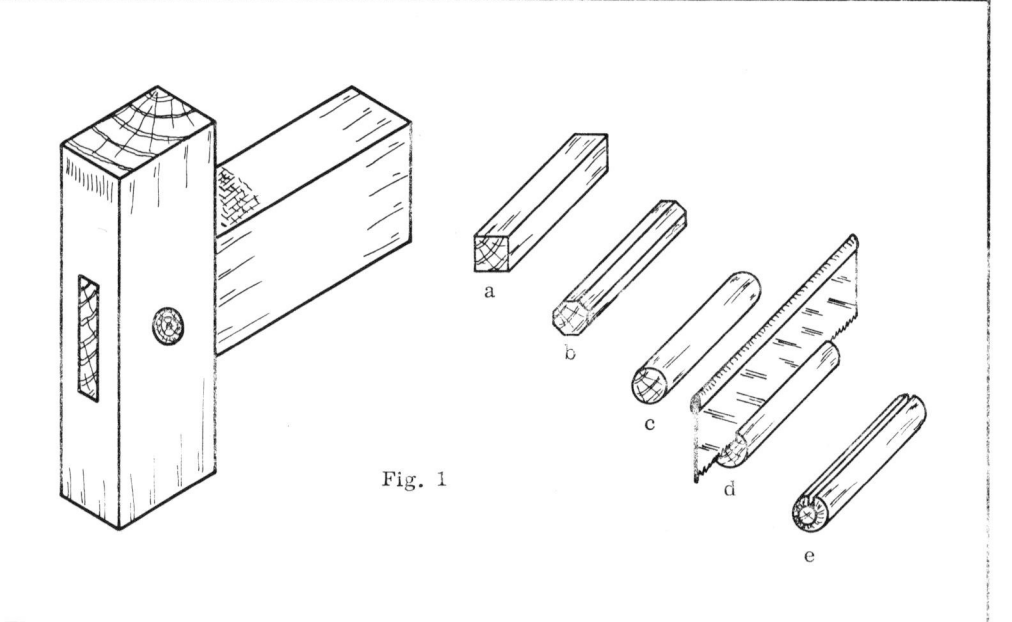

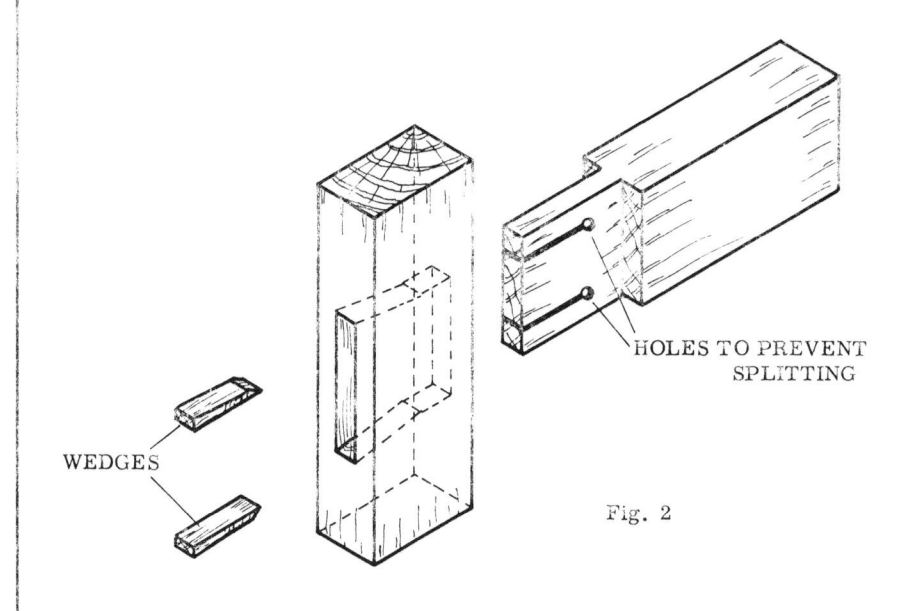

NOTES:

N P V C		FRAMING JOINTS.
123	BASIC.	

SECURING THE JOINTS

Instead of nails to secure mortice and tenon joints, either pegs or wedges can be used.

One or two holes are drilled through the assembled joint and wooden dowels, or pegs, as they are called in this case, are inserted with glue to securely fix the joint (Fig. 1).

- To make the dowels, plane off the corners of a square piece of hard wood, until the piece is round. When the dowel is cut to length, chamfer the ends and cut a groove along the length to permit air and excess glue to escape (Fig. 1, a - e).

Follow the steps below to secure a joint by means of wedges.

- Cut the mortice with an allowance of 2 mm in width, tapering from the outside edge to about 2/3rd of its depth (Fig. 2).
- Make cuts in the tenon to receive the wedges.
- To prevent splitting of the tenon, drill small holes at the end of each cut.
- Cut the wedges from small pieces of waste wood; they should have the same length as the tenon.

Haunched mortice and tenon joints in frame-like constructions should not be wedged, because of the danger of breaking off the small haunch at the corner of the joint.

Both wedges and pegs can be used for securing mortice and tenon joints in box-like constructions.

NOTES:

PIN

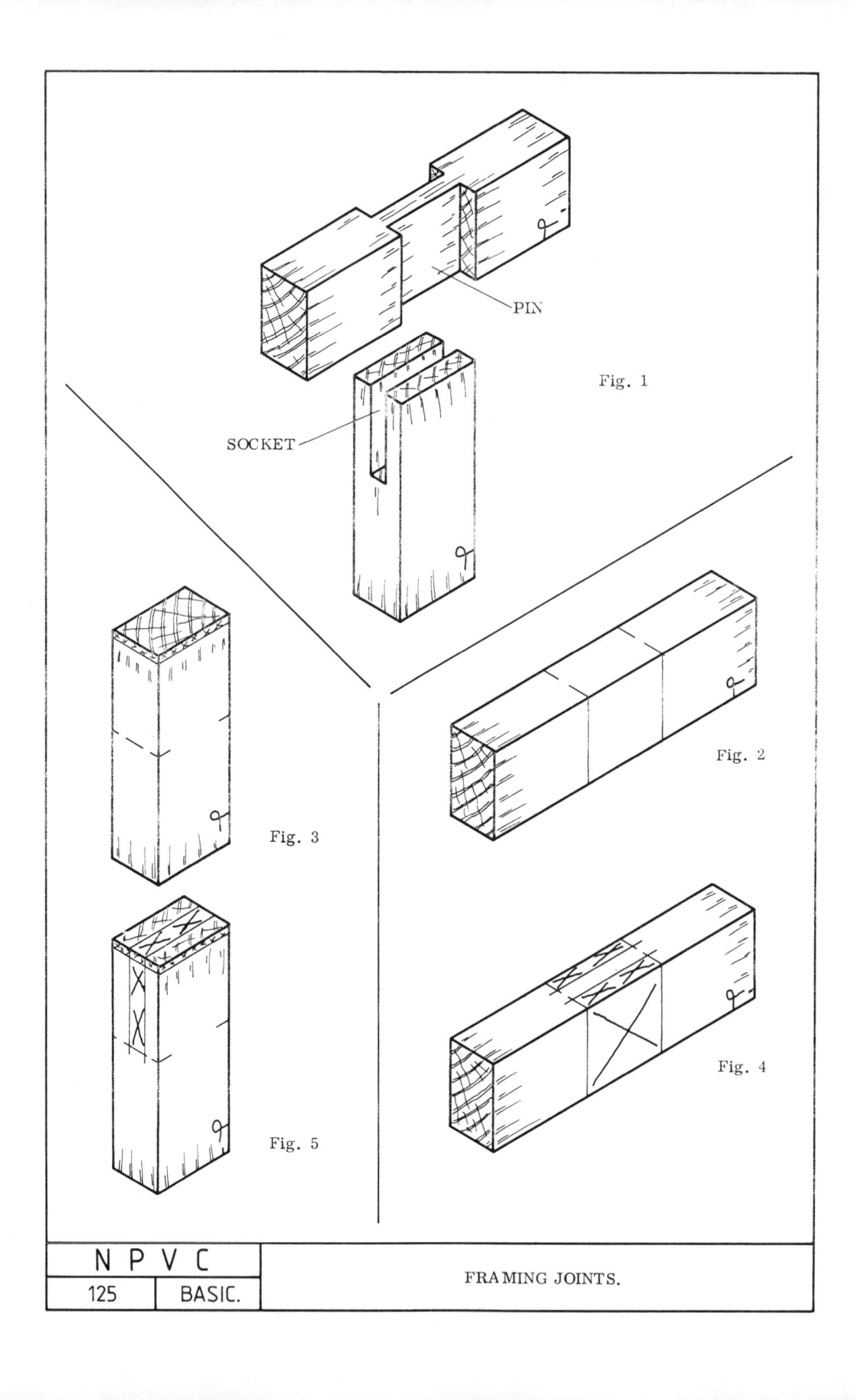

SOCKET

Fig. 1

Fig. 2

Fig. 3

Fig. 4

Fig. 5

N P V C		FRAMING JOINTS.
125	BASIC.	

BRIDLE JOINT

Bridle joints are similar to mortice and tenon joints. They consist of a pin and a socket (Fig. 1). The thickness of the pin is 1/3rd of the thickness of the member.

The two types of bridle joint are the tee bridle (Fig. 1) and the corner bridle. Here we will only go into detail about the tee bridle, since the construction of the corner bridle joint follows much the same procedure.

SEQUENCE OF OPERATIONS FOR CONSTRUCTING THE JOINT

Step 1. Preparation of the timber

- Make a cutting list.
- Prepare the timber.

Step 2. Marking out

- Mark the position of the pin on one member, making the distance between the shoulders equal to the width of the other piece. Square the lines all around the piece with a try square and pencil (Fig. 2).

- Mark the length of the socket (plus 2 mm waste) on the end of the other member, making the length equal to the width of the pin. Square the lines across the face side and on both edges (Fig. 3). Remember to smooth the pieces before using them to mark.

- Set a marking gauge to 1/3rd of the thickness of the member and gauge along both edges of the pin. Use the gauge from the face side only. Mark the waste with small crosses (Fig. 4).

- With the same setting on the gauge, mark around the end of the socket. Mark the waste (Fig. 5).

- Mark the other side of the socket in the same manner, from the face side, with the gauge set at 2/3rds of the thickness of the piece. If you have a gauge with 2 pins, mark both lines at once.

- Check the fitting.

NOTES:

FRAMING JOINTS.

Fig. 6

Fig. 7

Fig. 8

Fig. 9

Fig. 10

Fig. 11

SOCKET

PIN

Fig. 12

N P V C		FRAMING JOINTS.
127	BASIC.	

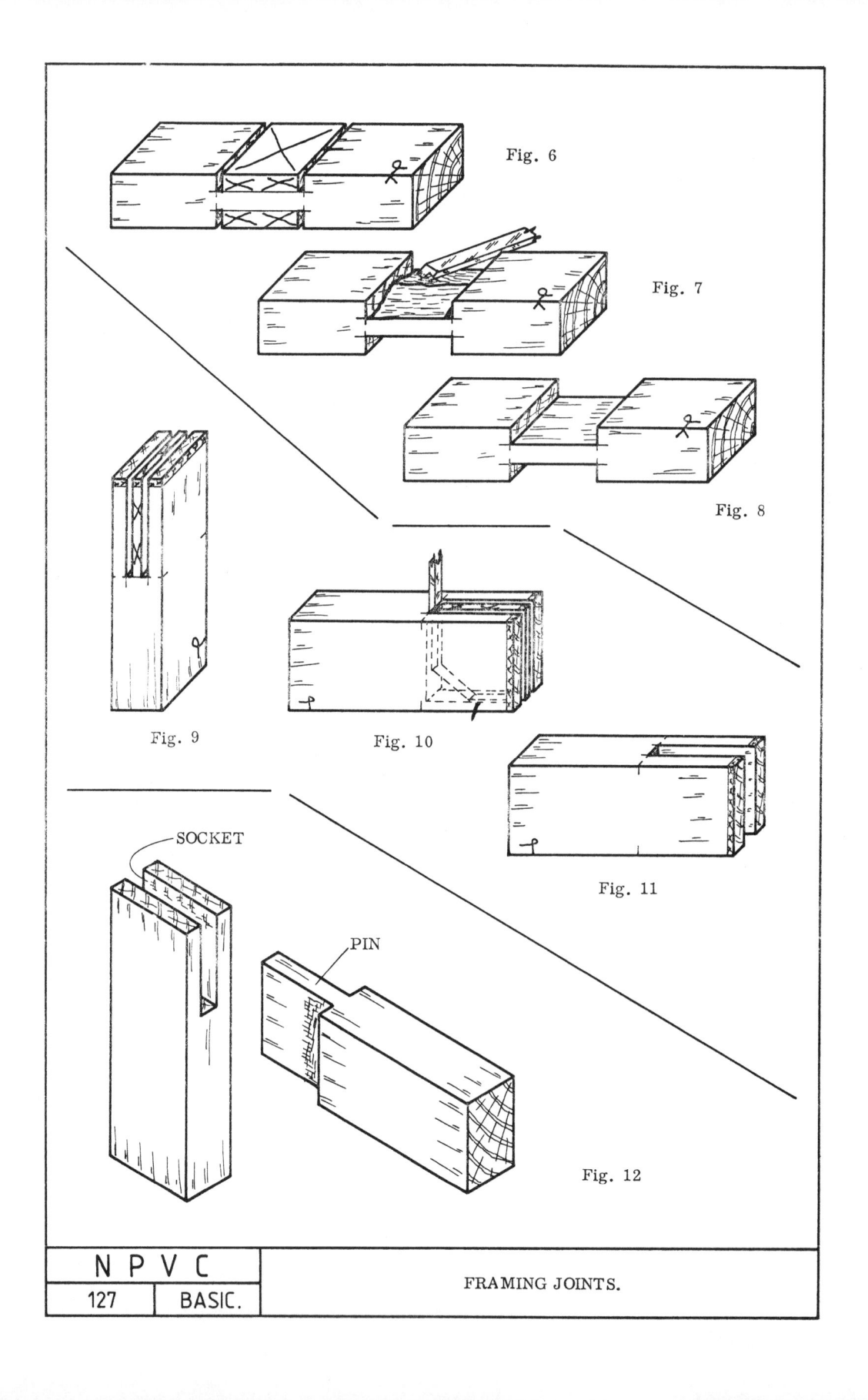

Step 3. Cutting the pin

- Carefully saw the shoulders down to the gauge line, sawing on the waste side of the line (Fig. 6).

- Chisel away the waste, chiselling halfway through from both edges (Fig. 7).

Step 4. Cutting the socket

- Rip the sides of the socket down to the required depth, sawing on the waste side of the lines (Fig. 9). Saw in steps (see Tee-halved joint, Cutting the pin, page 114).

- Chop out the waste with a mortice chisel, chiselling halfway through from both edges (Figs. 10 & 11).

Step 5. Assembling the joint

- Clean up the inside edges which cannot be reached after the joint is assembled.

- Assemble the joint with glue and nails.

- When the glue is dry, plane off the waste of the socket.

- Clean up the sides and edges with a smoothing plane.

CORNER BRIDLE JOINT

The corner bridle joint is used where members meet to form the corner of a frame. Like the tee bridle, it consists of a pin and a socket (Fig. 12).

The pin is constructed like the tenon in the sequence of operations for the mortice and tenon joint for frame-like constructions, pages 118 to 120.

The socket is constructed in the same way as the socket for the tee bridle joint, above.

NOTES:

Fig. 1

Fig. 2

DOWELS

Fig. 3

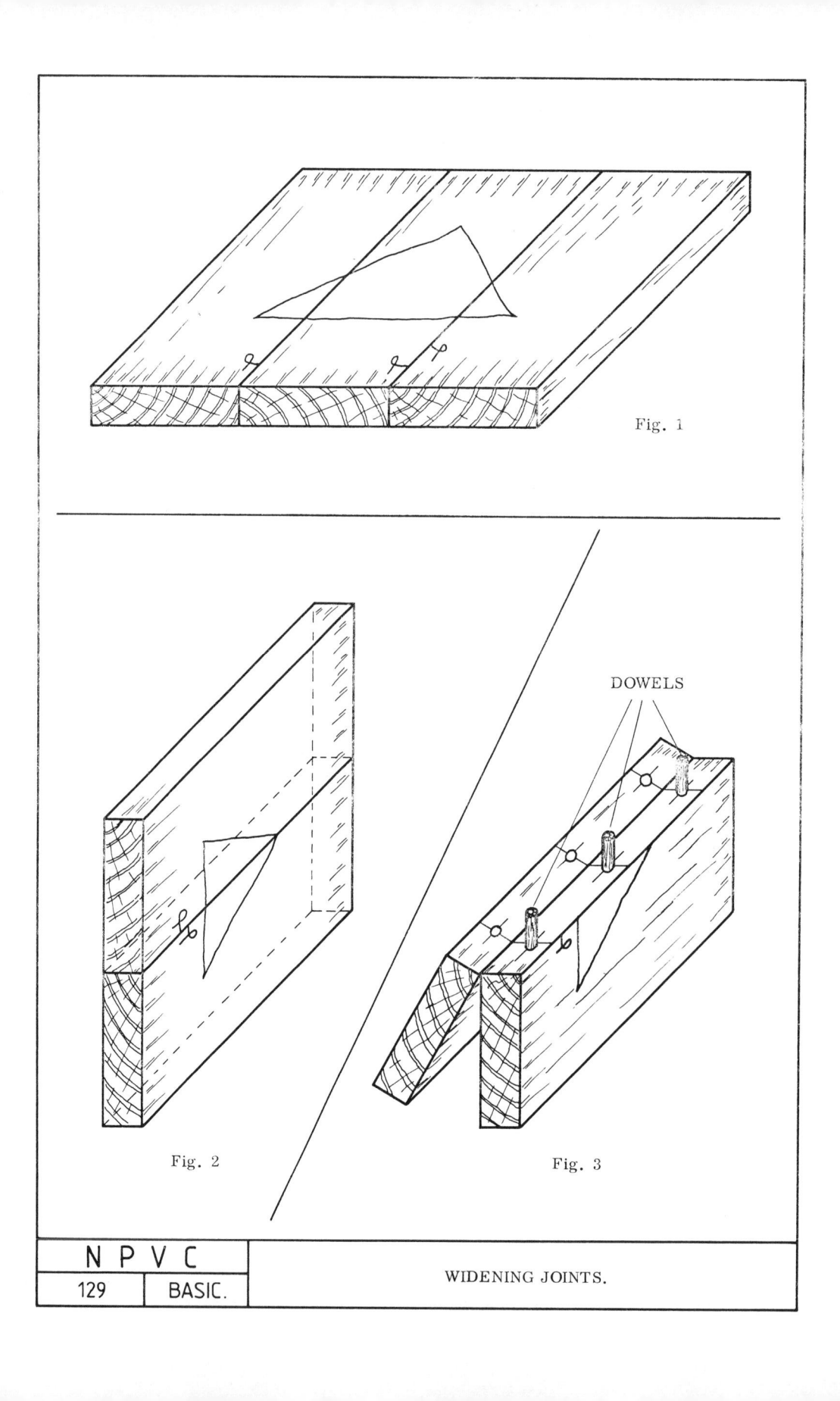

| N P V C | WIDENING JOINTS. |
| 129 BASIC. | |

WIDENING JOINTS

Widening joints are joints used to make a single, wide board by joining two or more narrow boards along their length, edge to edge (Fig. 1).

The boards that will be joined must first be marked. Lay the boards out in the desired position and mark them with a triangular mark over all the boards (Fig. 1). The triangle should point upwards. This mark will help us to keep in mind the position of each board during the steps that follow.

PLAIN GLUED BUTT JOINT

This is the simplest widening joint (Fig. 2). The edges of the boards are planed perfectly straight and square, and then butted together. The joint is glued and clamped tightly to force out the surplus glue. For narrow pieces this is done with G-clamps. For wider pieces, wooden or metal sash clamps are used.

DOWELLED WIDENING JOINT

This joint is similar to the plain glued butt joint, but strength is added by means of cylindrical wooden pins, called dowels. Dowels are made as explained in the section on securing joints. The dowels are then glued into holes in the edge of each board (Fig. 3). The diameter of the dowels should be about one-third of the thickness of the pieces that are being joined.

The holes should be about as deep as the boards are thick, and they should be slightly countersunk (see Fastening with Screws, page 96).

Mark out the position of the dowels by putting the boards on top of each other, sides together and marking both edges at the same time. The centre can be marked with a marking gauge, marking from the face side.

Metal or wooden sash clamps are used to press the boards together during glueing.

NOTES:

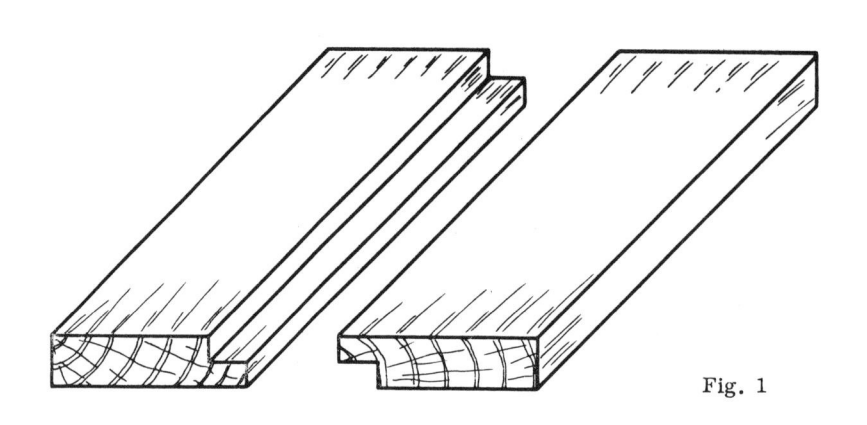

Fig. 1

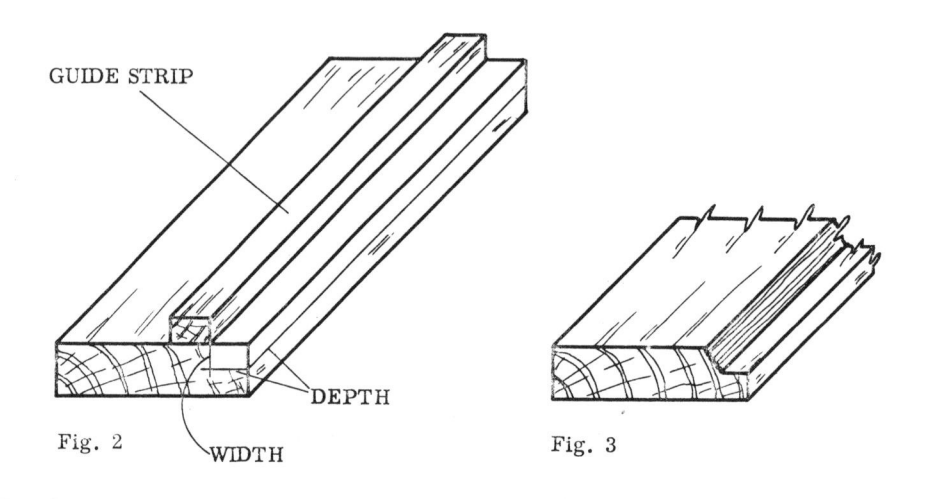

GUIDE STRIP

DEPTH

WIDTH

Fig. 2

Fig. 3

NOTES:

N P V C		WIDENING JOINTS.
131	BASIC.	

REBATED JOINT

In this widening joint, the edges of the boards are rebated to match each other (Fig. 1). The rebating is done with either an ordinary rebate plane or an adjustable one. This joint is stronger than the plain glued butt joint.

HOW TO PLANE A REBATE WITH AN ORDINARY REBATE PLANE

Step 1.
- Mark the depth and width of the rebate with a marking gauge (Fig. 2).

Step 2.
- Fix a wooden guide strip along the line that marks the width of the rebate (Fig. 2). The guide strip must be perfectly square and it should lie flat.

Step 3.
- Plane until you reach the line marking the depth of the rebate. Take care that the side of the plane is always against the guide strip, so that the width of the rebate is the same along the whole length.

- If you notice that you are planing against the grain, stop just before you reach the required depth and plane from the other direction. This will ensure that the surface of the rebate is smooth.

An important point in planing rebates is setting the plane correctly. The side of the cutting iron that faces the rebate must be set so it is exactly flush with or only slightly coming out at the side of the plane. If it projects too far it will damage the guide strip, and if it is set in from the side it will not plane true (Fig. 3).

When you set the cutting iron, do not knock on it with a steel hammer. This will damage the iron. Rather, loosen the wedge slightly and knock it with a mallet or a piece of wood.

When the rebate plane is not set well, it will tend to slip off the rebate and will not produce a good surface.

NOTES:

DEPTH GUIDE (A)

WIDTH GUIDE (B)

Fig. 1

Fig. 2

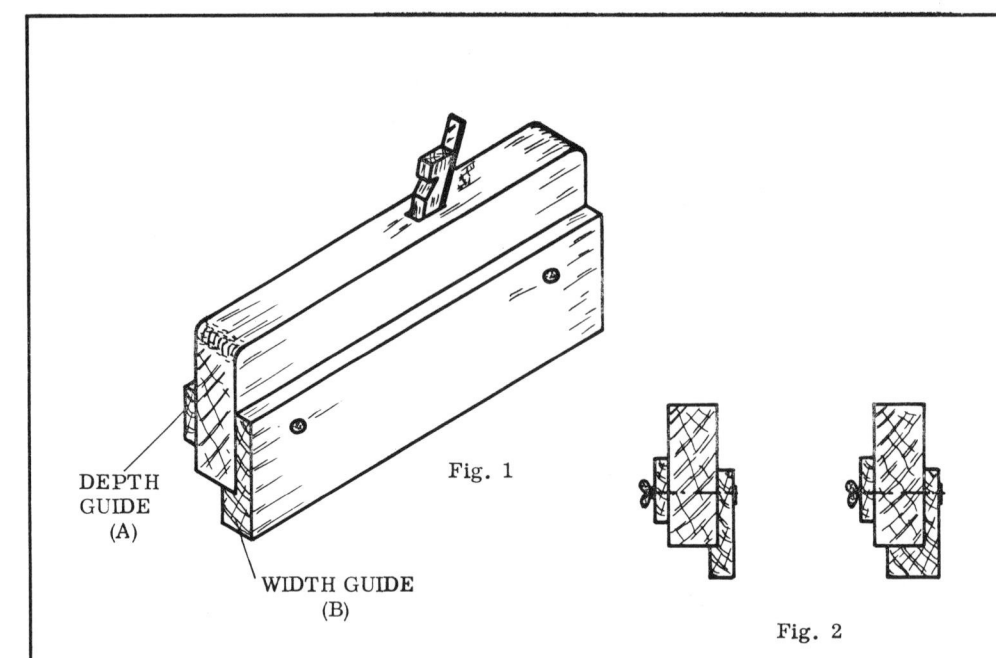

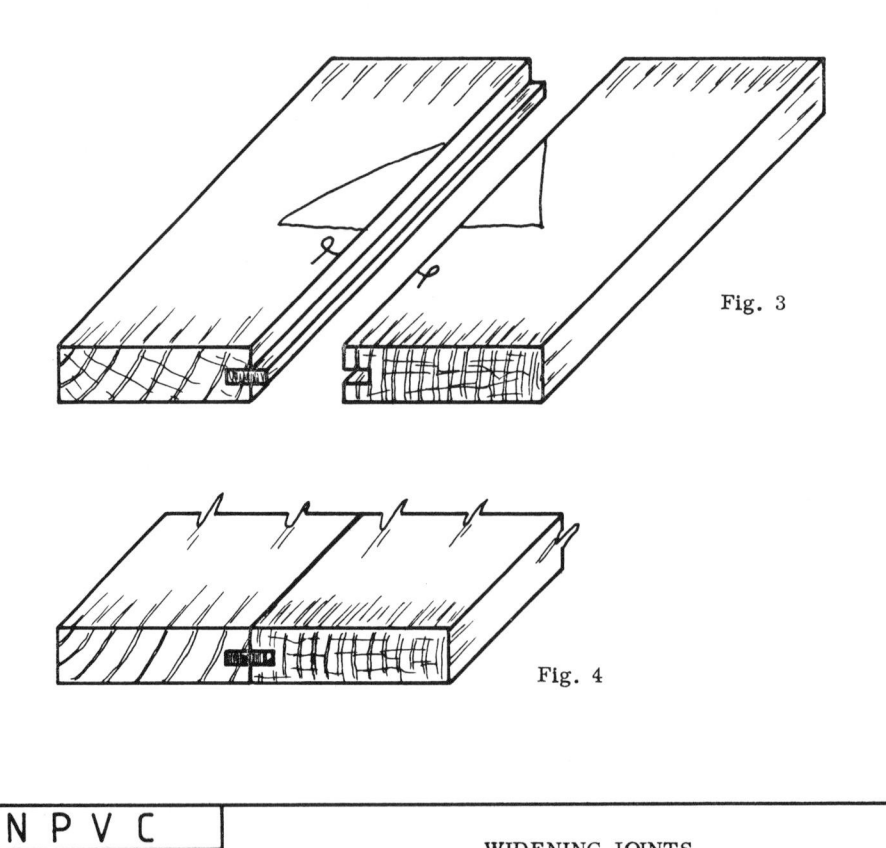

Fig. 3

Fig. 4

N P V C		WIDENING JOINTS.
133	BASIC.	

HOW TO PLANE A REBATE WITH AN ADJUSTABLE REBATE PLANE

To make work simpler, we can fix guides onto the rebate plane itself. Thus, fixing guide strips on the boards is unnecessary (Fig. 1).

One wooden piece is fixed on the sole of the plane (B) at the standard width for rebates and another piece is fixed on the side of the plane (A) and can be moved up or down to adjust the depth of the rebate. The width can also be adjusted, by using a wider or narrower wood guide (Fig. 2).

The guides should not be nailed to the plane, since that would damage it. They should be fixed by bolts and nuts, so that they can be easily removed.

Plane until the depth guide just touches the work. Take care that the width guide is always firmly pressed to the side of the timber. If you notice that you are planing against the grain, stop just before you reach the required depth and finish planing with the guide strips removed, which enables you to plane in the other direction. This gives a good surface to the rebate.

See the section on the ordinary rebate plane for tips on how to set the cutting iron.

LOOSE TONGUED JOINT

This joint is used where a joint stronger than the plain glued or rebated joint is needed. The boards to be joined must be at least 2 cm thick (Fig. 3).

The joining edges are grooved and a tongue is glued into the grooves. The depth of the groove is about 2/3rd of the thickness of the board. The width of the groove is equal to the thickness of the tongue. The groove should be slightly deeper than the projection of the tongue, to allow for expansion (Fig. 4).

Plywood makes a very strong tongue and it is frequently used for this purpose.

If solid wood is used as a tongue, care must be taken that it is always cut across the grain. A tongue cut with the grain will make a weak joint.

NOTES:

PROTECTIVE COVER
TO BE FITTED

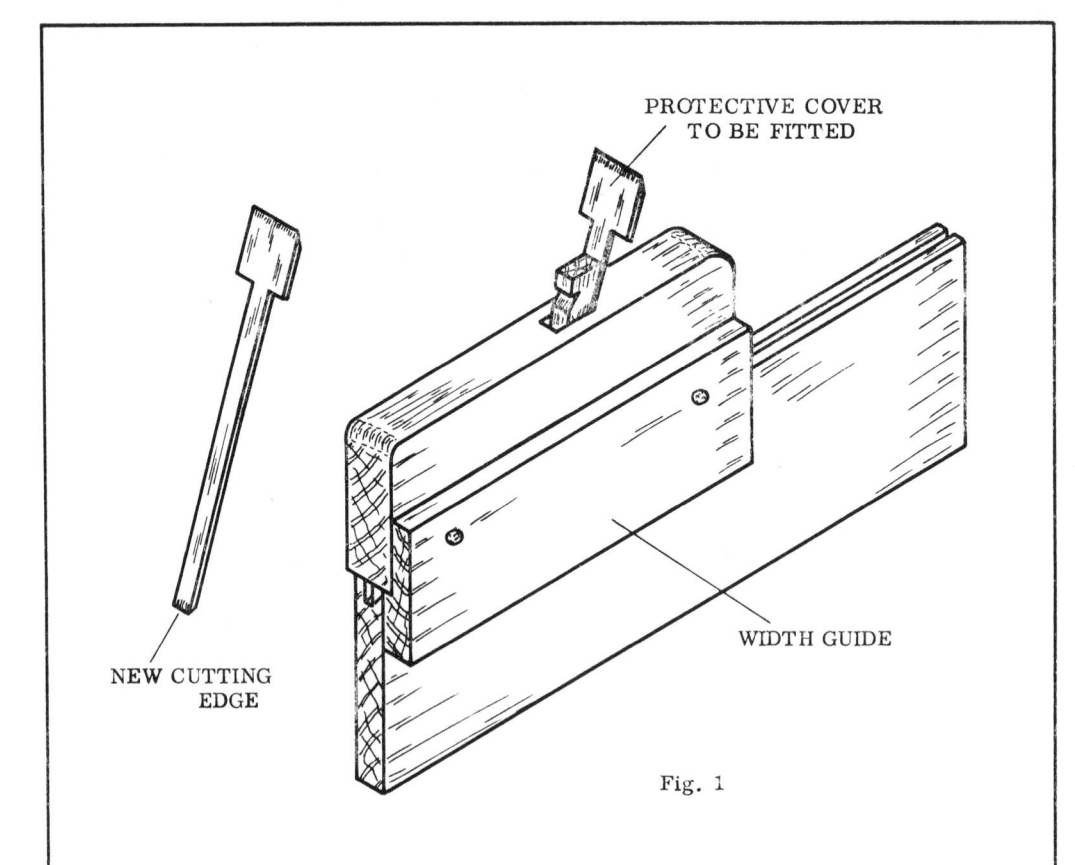

NEW CUTTING
EDGE

WIDTH GUIDE

Fig. 1

NOTES:

N P V C		
135	BASIC.	WIDENING JOINTS.

HOW TO PLANE A GROOVE FOR A LOOSE TONGUED JOINT

Usually special planes called plough planes are used to plane grooves for this kind of joint. If a plough plane is not available, we can adapt our rebate plane for this purpose and make an improvised plough plane (Fig. 1).

To do this, grind and sharpen the narrow end of the rebate plane cutting iron to make a cutting edge. Grind the sides of the iron to the size of the most commonly used tongue, which is 6 mm plywood. The sides should be slightly bevelled to ensure free movement in the groove.

When the iron is fitted into the plane it is adjusted so that the cutting edge projects out of the sole by exactly the required depth of the groove.

A guide, similar to the one used for the adjustable rebate plane, is now fitted to the side of the plane. This guide keeps the cutting iron at the right distance from the face side of the boards. It should be adjusted according to the most common thickness of the boards, in this case it is about 22 mm for a planed board. For a tongue size of 6 mm then, the distance between the edge of the guide and the edge of the cutting iron will be 8 mm.

When planing press the guide firmly against the side of the wood and hold the plane exactly at a right angle to the edge of the board.

The most difficult part will be to start the groove, since the cutting iron will tend to slip off the edge and it requires some experience to keep it steady. Go slowly at first.

Work from the face side at all times.

To prevent injuries cover the cutting edge where it sticks out of the top of the plane.

NOTES:

Fig. 1

Fig. 1a

Fig. 2

Fig. 3

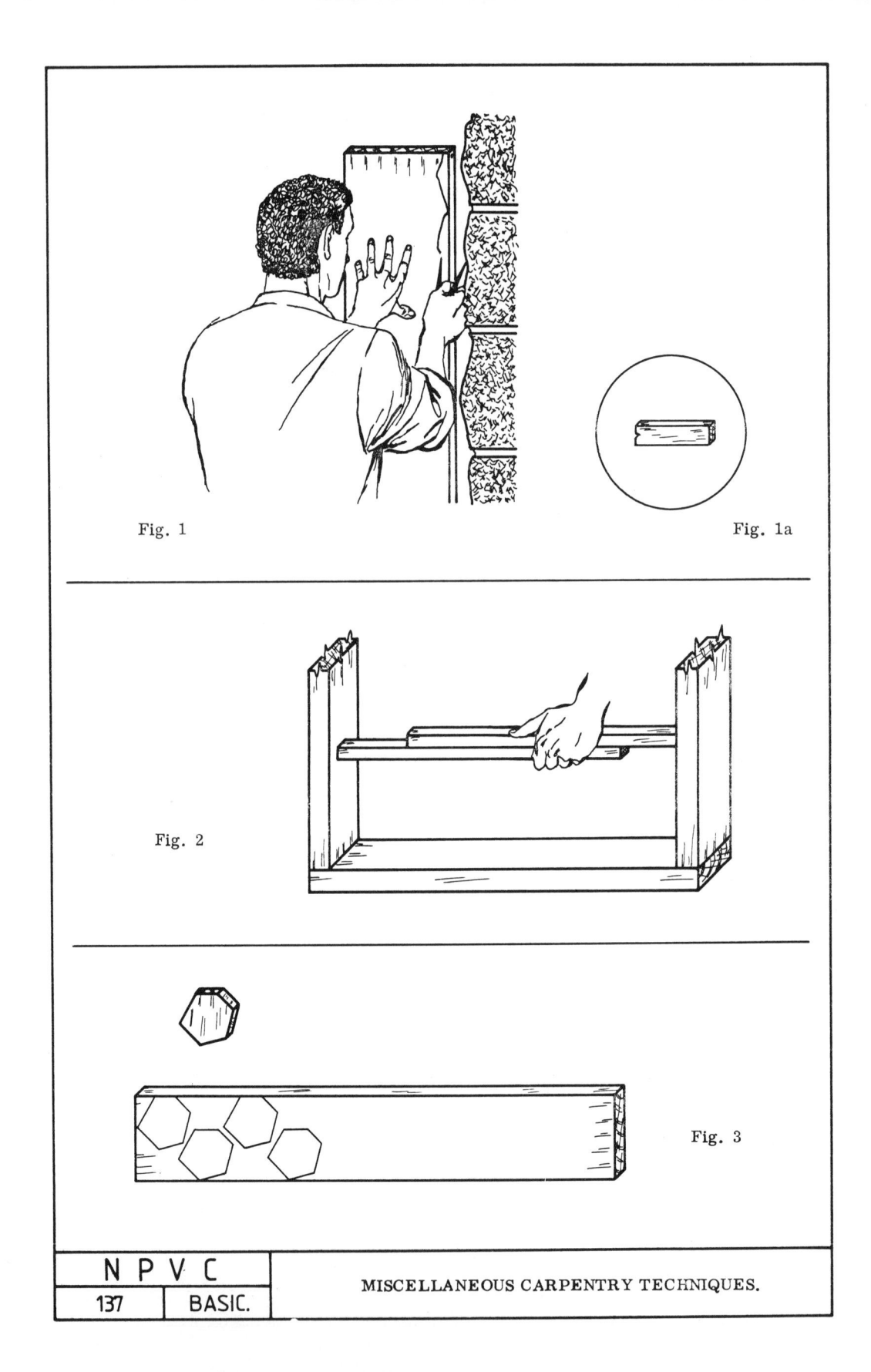

NPVC		MISCELLANEOUS CARPENTRY TECHNIQUES.
137	BASIC.	

MISCELLANEOUS CARPENTRY TECHNIQUES

MARKING A BOARD TO FIT AN IRREGULAR SURFACE

To mark the edge of a board which you want to fit against an irregular surface such as an unplastered wall, hold the board firmly and level to the wall and mark it with a compass or a similar device as shown in Fig. 1.

As one leg of the compass moves along the wall, the other leg will mark on the board an exact copy of the irregularities of the wall surface. The legs of the compass have to be set apart by a distance a little greater than the width of the biggest gap between the wall and the board.

If no compass is available, a small wooden block can be used instead (Fig. 1a). The pencil is held in the notch at one end and the other end is moved along the surface of the wall.

MEASURING THE WIDTH OF OPENINGS

A convenient way to measure the width of openings such as doors and windows is to use two sticks as shown in the illustration (Fig. 2) to just span the opening. Then transfer the measurement to a single board by marking and measure it with a rule.

MARKING OUT IRREGULAR DESIGNS WITH TEMPLATES

When you want to mark out several pieces with the same irregular shape, you can save time and ensure more accurate work by marking from a template (Fig. 3). Templates are thin pieces of cardboard or plywood onto which the required pattern is drawn and then cut out.

The template is used by placing it on the material to be marked and holding it firmly in place while drawing around it.

NOTES:

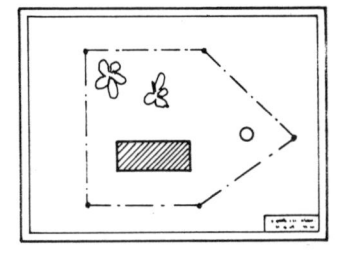

Fig. 1
LOCATION PLAN

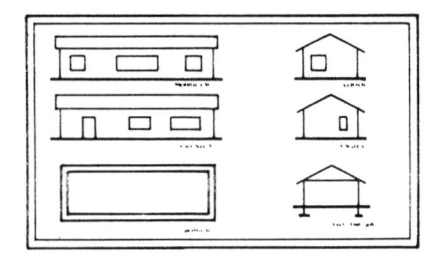

Fig. 2
BUILDING PLAN

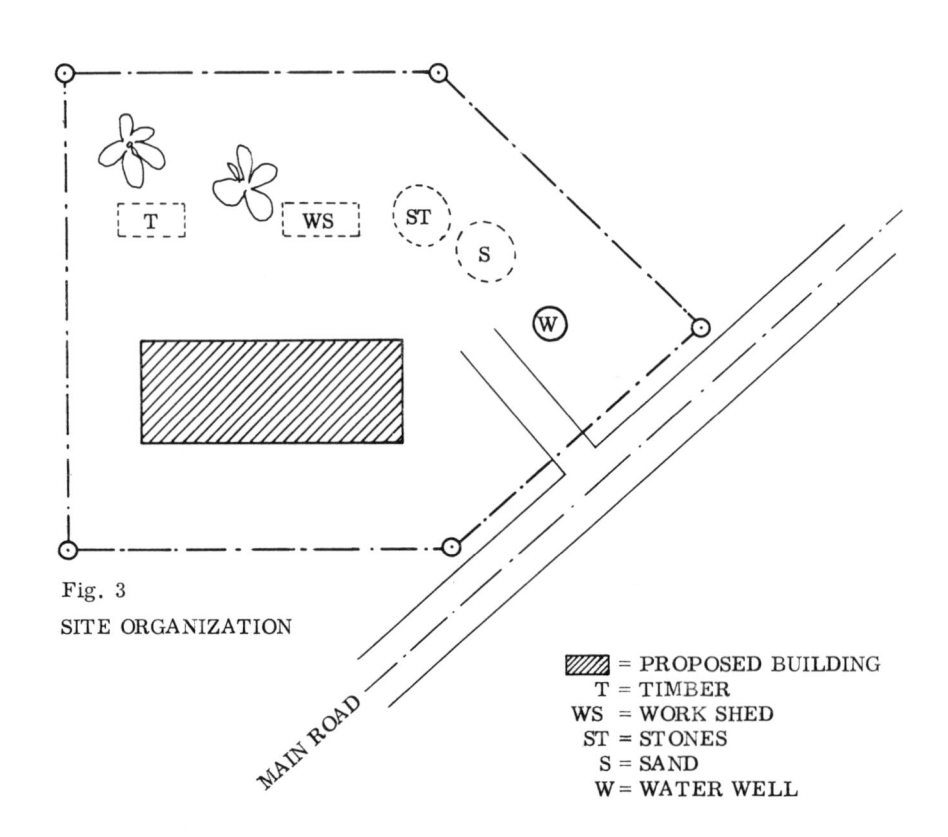

Fig. 3
SITE ORGANIZATION

MAIN ROAD

▨ = PROPOSED BUILDING
T = TIMBER
WS = WORK SHED
ST = STONES
S = SAND
W = WATER WELL

NPVC	PART 3: PREPARATION FOR ON-THE-JOB TRAINING.
139	JOB.

BUILDING PRELIMINARIES.

PART 3: PREPARATION FOR ON-THE-JOB TRAINING

BUILDING PRELIMINARIES

Before anyone can actually start to erect a building, a number of preliminary steps must be completed. The very first step is the preparation of the plan.

PLAN

The plan, also called the drawing, is a layout of a building drawn on paper. It contains all the information necessary to erect the house (see Drawing Book, page 55). The data and measurements given in the plan are essential for the builder to be able to construct the building so that it satisfies the customer's demands (Figs. 1 & 2).

PLOT AND SITE CLEARING

A plot is an area of land containing one or more sites. It is determined and limited by boundaries (Fig. 1).

The site is that area of land within the plot which is actually used for construction.

With the location plan in hand, the builder can prepare both plot and site for the construction of the building. The location plan tells him exactly where the trees and bushes have to be removed so that they don't interfere with the work. This preparation includes making a drive, cutting the grass, and levelling the surface of the ground.

The builder must pay special attention to the roots of trees which are on the site or very close to it. These must be completely removed. If some roots, such as those of the neem tree, remain in the ground, they can grow again and damage the structure.

SITE ORGANIZATION

When the land clearing is completed, the building materials can be brought in to the building site.

Temporary work-sheds and stores may be erected in suitable places (Fig. 3).

The builder must ensure that there is an adequate supply of clean water. Without water no building can be constructed.

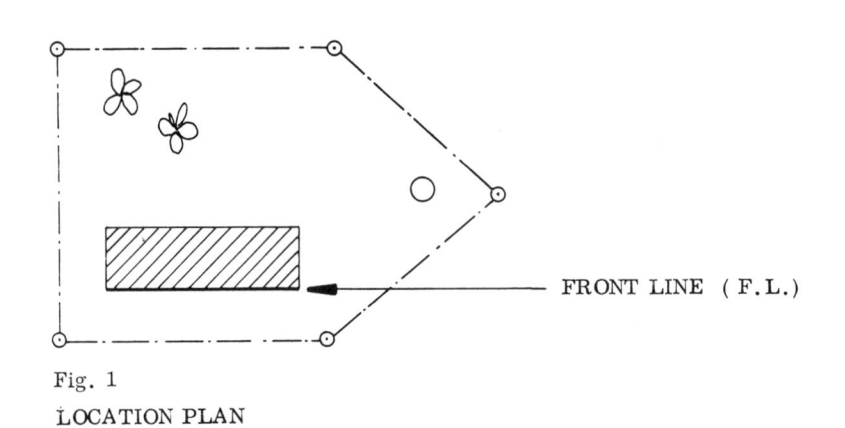

Fig. 1 LOCATION PLAN

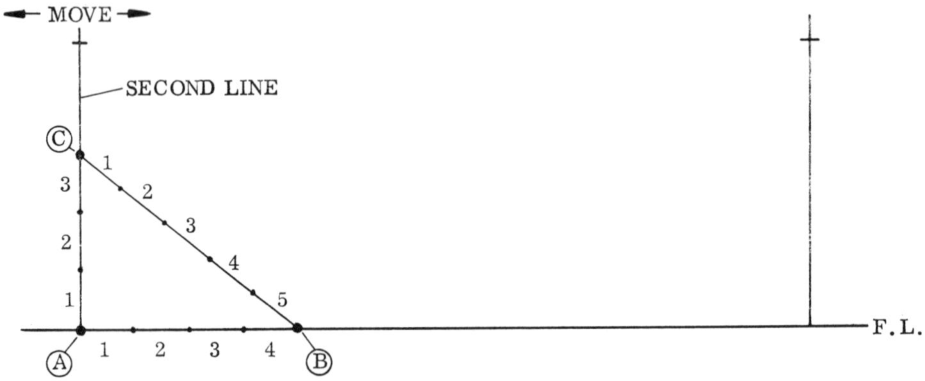

Fig. 2 3-4-5 METHOD

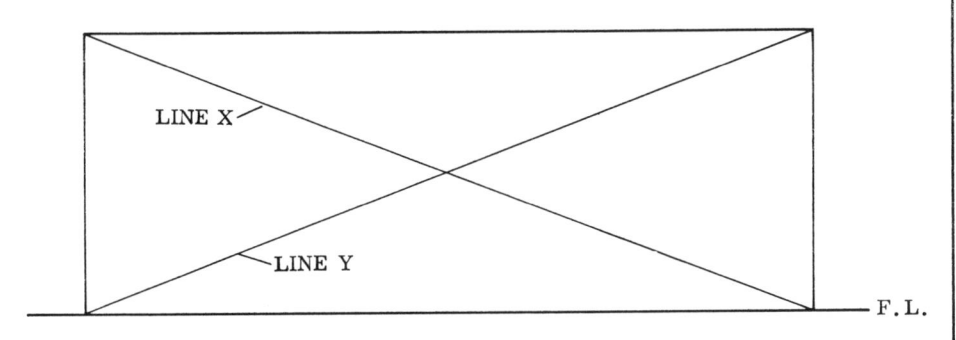

Fig. 3 LINES X AND Y ARE EQUAL IN LENGTH

N P V C
141   JOB.

SETTING OUT.

SETTING OUT

At the beginning of any construction activity the work must be carefully set out. This is also known as pegging out or lining out.

Setting out means to put pegs in the ground to mark out an excavation; or to mark on the floors to locate walls.

3 - 4 - 5 METHOD

The first line to be set out is the front line of the carcass (Fig. 1). A "carcass" is the building when it is structurally complete but otherwise unfinished. In this case we mean that the front line marks the position of the outside face of the (future) unplastered wall. The lines of all the other walls are measured from this front line. If the building is rectangular, right angles are set off from the front line by using the 3-4-5 method.

The second line to be set out is the line of one of the side walls of the carcass. This line intersects the front line at the corner of the future building. To make sure that this corner is a right angle, we use the 3-4-5 method.

- Measure a distance of 4 m along the front line starting from point A, and mark this on the line (point B) (Fig. 2).

- Measure a distance of 3 m along the second line, starting from the corner (point A) and mark this distance (point C).

- Now take a line which is marked with a distance of 5 m, and stretch it taut from point B towards the line with point C. Keeping the end points of both lines steady (points A & B) and the lines taut; move the free ends of the side line and the 5 m line until the 5 m mark and the mark at point C meet each other. This is best done with two men, one at the end of each line.

- The corner angle must now be a right angle.

- Measure the required length of the side line and insert a peg at the end. Set out the opposite side line in the same way.

If the setting out has been done accurately, the length of the back line between the two pegs should be equal to that of the front line. Make a further check by measuring the diagonals, which must be equal (Fig. 3).

NOTES:

SETTING OUT.

N P V C
JOB. 142

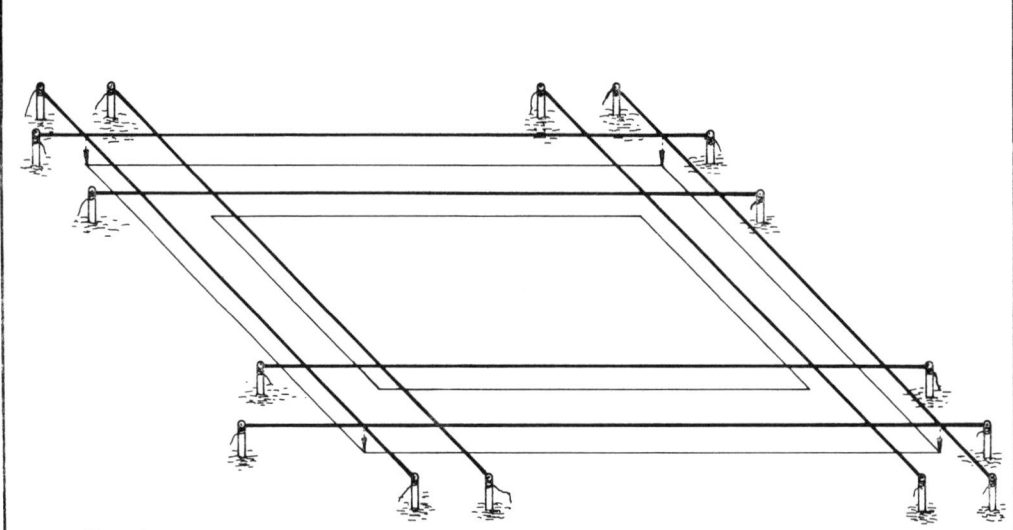

Fig. 1
SETTING OUT THE LINES & MARKING THE GROUND

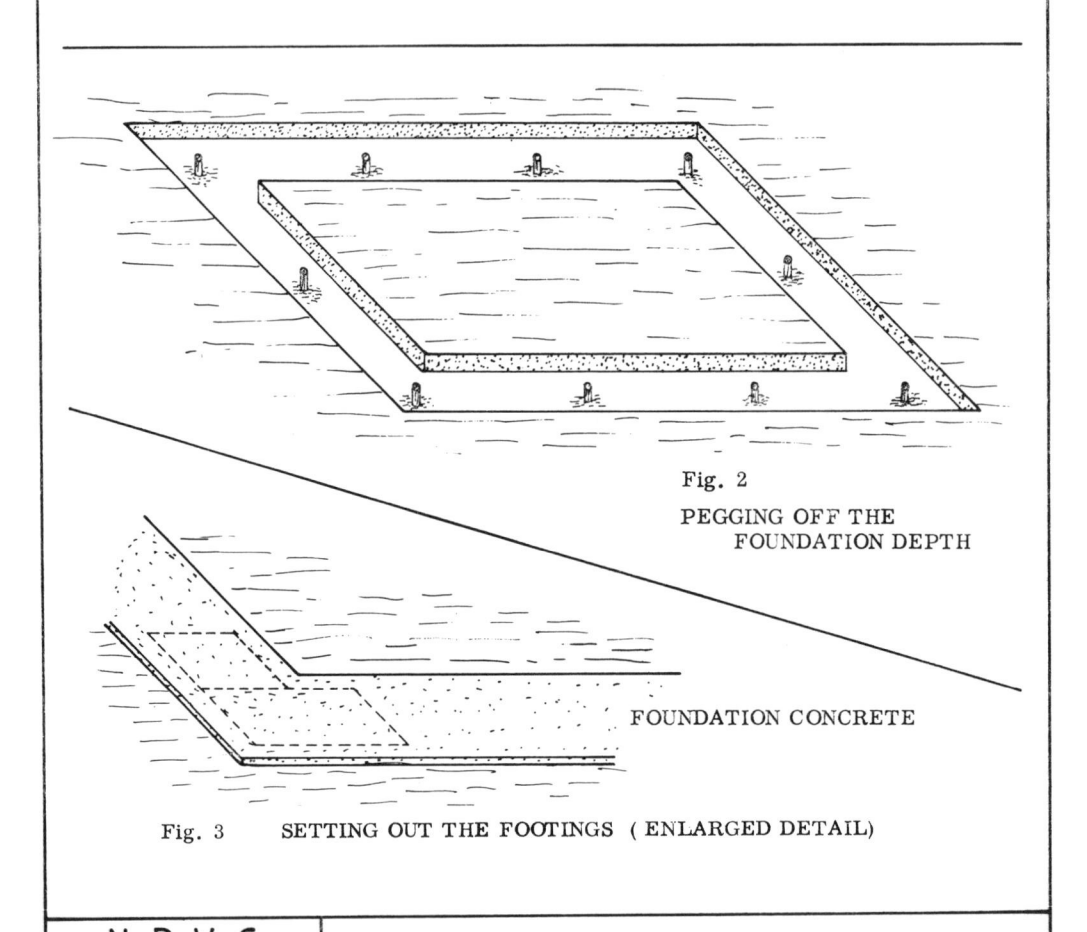

Fig. 2
PEGGING OFF THE FOUNDATION DEPTH

FOUNDATION CONCRETE

Fig. 3   SETTING OUT THE FOOTINGS (ENLARGED DETAIL)

NPVC		SETTING OUT.
143	JOB.	

LINING OUT

Once the positions of the corners and the distances between them are determined, the positions of the foundations, footings and walls as well as their thickness must be marked. A simple example of setting out and marking a foundation is shown in Fig. 1. The more complicated and permanent methods will be treated later.

These marks will be needed until the plinth course is completed, so they must be relatively durable, so that they remain accurate for a longer period and are not destroyed by rain or other influences.

DIRECT MARKING

Small buildings or small extensions of houses may be marked directly on the flat ground, provided that the excavation work can proceed immediately and can be quickly completed, so that the marking need not be repeated (Fig. 1).

In this procedure, the setting out must be done in stages.

- Mark the position and width of the foundation directly on the ground, and dig the trenches immediately.

- The next step is to level the bottom of the trenches and to peg off the foundation depth (Fig. 2).

- After the foundation concrete is cast and set hard, set out the footings directly on the surface of the foundation (Fig. 3) and build them to the required height.

- When the footings and hardcore filling are complete, set out the plinth course on the footings.

- NOTE: The information given in this section and in most of the following sections is not intended to be a detailed explanation. It is simply meant to give you, the trainee, an idea of the operations you can expect to encounter in your on-the-job training. Lining out, marking, etc. will be covered in detail in the Construction Book.

NOTES:

SETTING OUT.   N P V C   JOB. 144

Fig. 1
USING THE PLUMB BOB

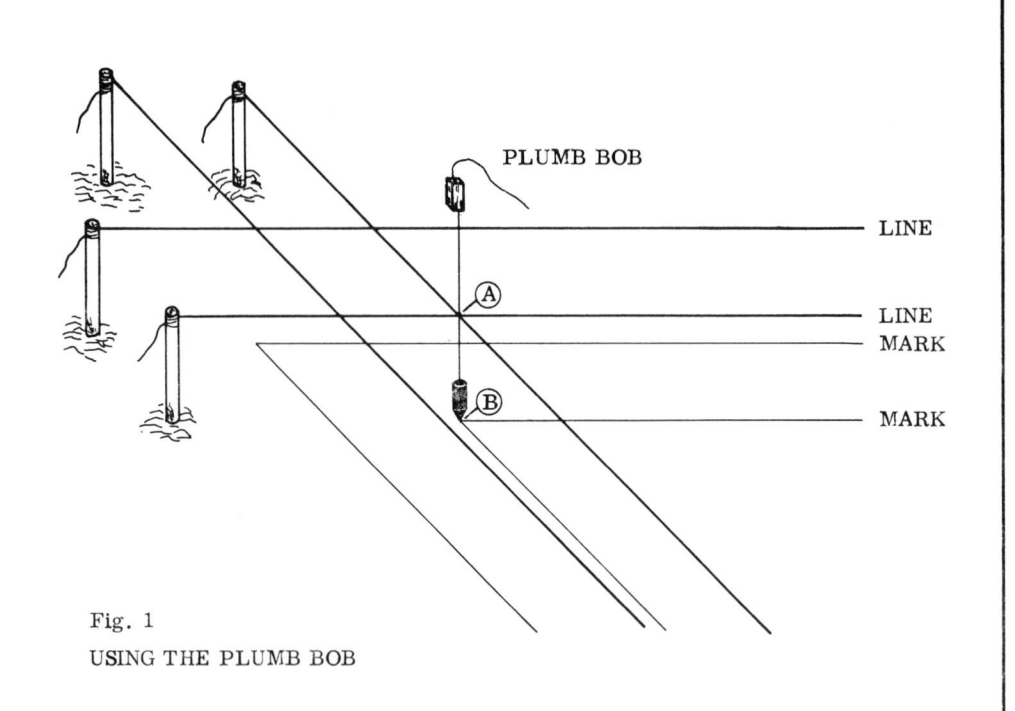

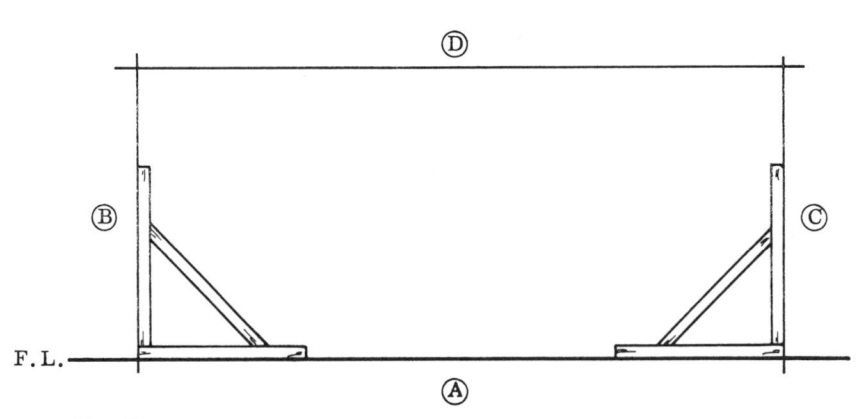

Fig. 2
USING THE LARGE SQUARE

PROCEDURE:

1 - FIX THE FRONT LINE; LINE A.
2 - MARK POINTS 1 AND 2.
3 - SET OUT LINE B SQUARE TO LINE A.
4 - SET OUT LINE C SQUARE TO LINE A.
5 - LINE D SHOULD BE THE SAME LENGTH AS LINE A.

N P V C		SETTING OUT.
145	JOB.	

USING THE PLUMB BOB TO MARK THE FOUNDATIONS

Hold the plumb bob with one hand by the suspending line so that the tip of the cylinder is just off the ground. Move it slowly until the suspending line just touches the intersection of the lines stretched between the pegs (see A & B, Fig. 1). When the swinging movement of the plumb bob has stopped, mark the point directly below the tip of the cylinder by inserting a peg.

The peg is directly in line with the intersection of the lines above.

This procedure is repeated at all inside corners and outside corners, so that the edges of the foundation trenches can be marked on the ground.

USING THE LARGE SQUARE

The large square, described in the Reference Book, Tools, page 12, may be used to set out and mark off the positions of inside walls. This is less time-consuming than using the 3-4-5 method.

Place the large square on the ground with one side along an already determined line, and mark off the corner on the other side (Fig. 2).

Not only the whole building, but also each room in the building must be checked for squareness by comparing the diagonals, which have to be equal.

USING THE MASON SQUARE

Although it is less accurate than the large square because of its smaller size, the mason square can be used to mark off the corners of short set-backs such as niches designed to receive built-in wardrobes, etc. Follow the same procedure as with the large square (Fig. 2).

- NOTE: A niche, also called a blocked doorway, is a small recess in a wall, usually not extending to the ceiling. A set-back or return is the part which goes back, away from the front or direct line of the structure.

NOTES:

SETTING OUT.

N P V C
JOB. 146

Fig. 1
CORNERS ARE NOT SHARP

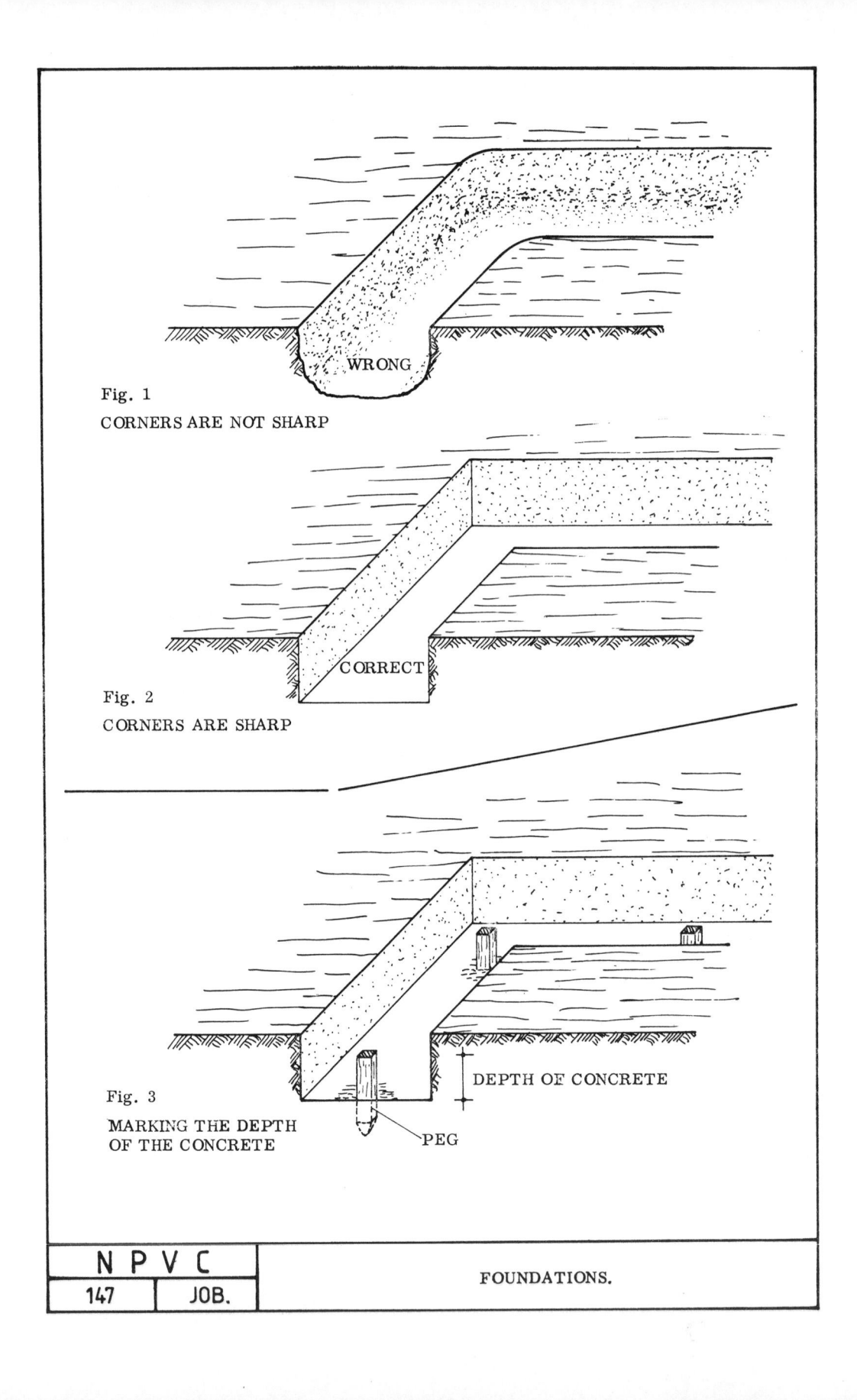

WRONG

Fig. 2
CORNERS ARE SHARP

CORRECT

Fig. 3
MARKING THE DEPTH
OF THE CONCRETE

DEPTH OF CONCRETE

PEG

NPVC	
147	JOB.

FOUNDATIONS.

FOUNDATIONS

A foundation is the strong base of a building. It is the lowest part of the structure, the part which is in direct contact with the ground.

The purpose of the foundation is to receive the loads from the structure above and to spread them over a larger area of supporting soil or rock.

EXCAVATING THE FOUNDATION TRENCHES

Once the setting out is completed and the position of the foundation is marked on the ground, the next step is to dig the trenches for the foundation concrete.

Remove the loose, soft topsoil to uncover the firm subsoil, preferably rocky soil. Dig the trenches to the required depth.

The soil which is taken out should be piled within or near to the area of the future building, so it can be used later for the hardcore filling. Take care to make the sides of the trenches vertical, and the bottom level. The corners should be sharp, not rounded (Figs. 1 & 2).

MARKING THE DEPTH OF THE CONCRETE AND LEVELLING THE TRENCH

When the trenches have been dug, the next step is to mark the depth of the foundation concrete. This is done by driving pegs in the bottom of the trench. The pegs are levelled across their tops, and their height above the trench bottom should be equal to the planned depth of the concrete bed (Fig. 3).

If the exposed heights of the levelled pegs are not equal, that means the bottom of the trench is not level. The uneven spots have to be levelled by taking out some soil, until all the pegs project equally.

- NOTE: Never level the bottom of a trench by adding loose soil, as this might lead to uneven settlement which causes cracks in the structure. Trenches are always levelled by removing soil. If this means that the trench is deeper than planned, then either the foundation concrete has to be deeper or the height of the footings must be increased.

NOTES:

FOUNDATION CONCRETE

When the trenches are dug and the thickness of the concrete has been marked, the next step is to mix the foundation concrete. The proportions for the mix can be from 1:10 to 1:15 (Reference Book, pages 166 to 170; and Tables of Figures, page 234).

CASTING - COMPACTING - LEVELLING

If the work is done during the dry season, the sides and bottom of the trenches must be watered down before the ready-mixed concrete is cast. This keeps the soil from absorbing too much moisture from the concrete before it has set. The concrete is carefully poured into the trenches and compacted by tamping.

Rammers are used to compact the concrete (Reference Book, page 18). The heavy rammer is repeatedly lifted and dropped, compacting (packing together) the comparatively stiff concrete.

A strike-board (Reference Book, page 25) is used to level the concrete to the height of the pegs. A straight edge can also be used instead of the strike board.

If wooden trench pegs were used, remove them now and fill in the holes.

If iron pegs were used, they can be left in the concrete unless they are needed for another job.

CURING

Cover the top of the freshly cast foundation with empty cement bags or straw. This keeps it from drying out in the sun and air, and keeps the surface clean.

Once the concrete starts to harden, the top of the foundation should be kept wet.

NOTES:

FOUNDATIONS.

FOOTINGS

The term footing is given to the courses of brickwork, stone or blockwork at the foot of a wall. The footing courses start immediately above the foundation and are laid flatwise. The rising wall is erected in the middle of the footing courses, so the footings, which are wider than the rising wall, project equally on both sides of the rising wall.

PURPOSE OF FOOTINGS

Two main functions must be fulfilled by the footings:

- They are the connecting link between walls and the foundation and act as an intermediate foundation for the walls, spreading the loads over a wider area of the concrete below.
- They raise the floor level high enough above ground level to keep water out during the wet season.

HEIGHT OF FOOTINGS

In Rural Building, the top of the foundations is usually at ground level, although they can be either above or below ground level depending on the subsoil.

When three footing courses are laid on top of the foundations, the soffit of the future floors will be 51 cm above ground level. This will meet the requirements of most situations.

If the building is in a valley, or in a place where the rain-water cannot run off quickly, the height of the footing courses must be increased.

NOTES:

Fig. 1
FOOTING COMPLETED

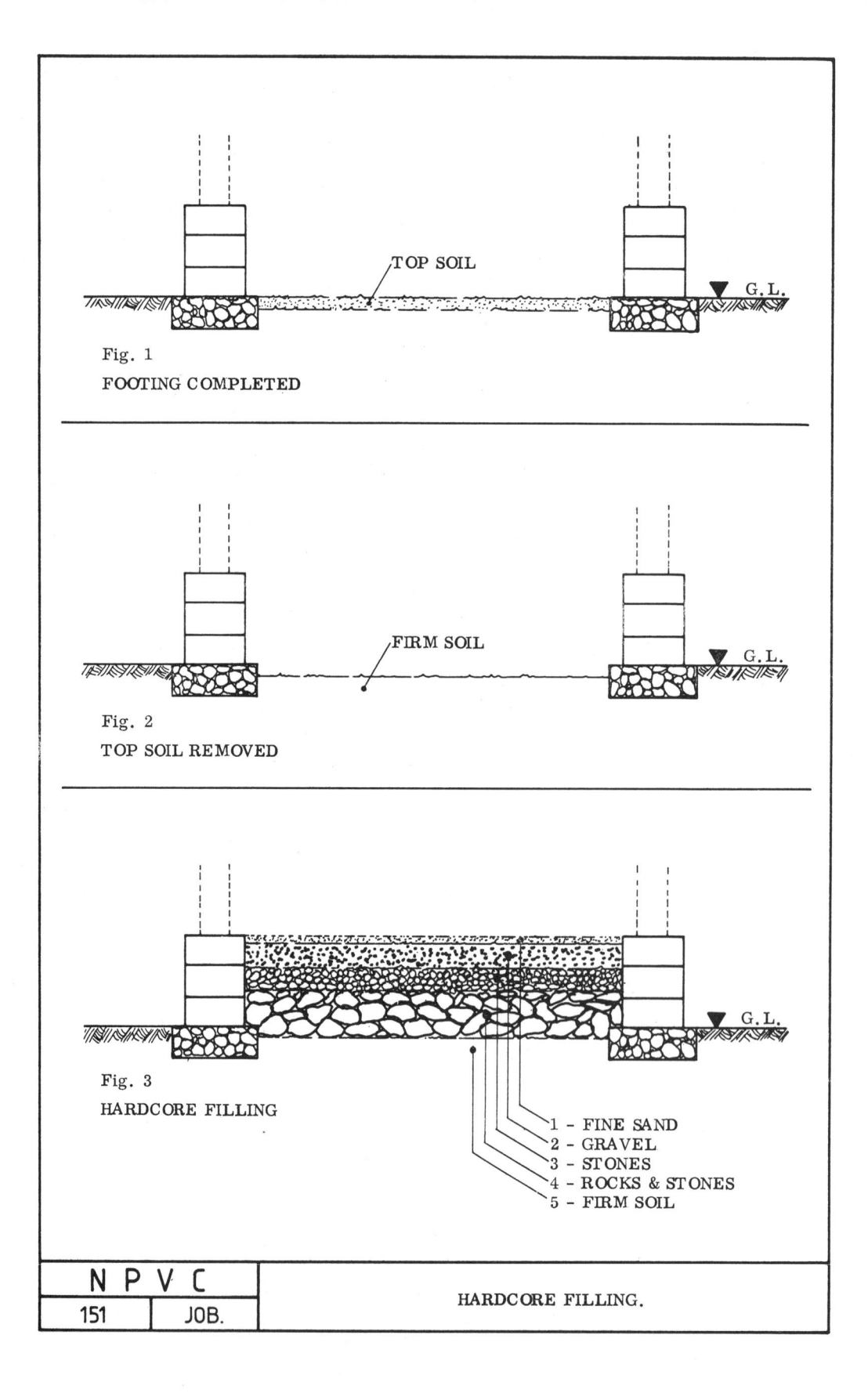

Fig. 2
TOP SOIL REMOVED

Fig. 3
HARDCORE FILLING

1 - FINE SAND
2 - GRAVEL
3 - STONES
4 - ROCKS & STONES
5 - FIRM SOIL

N P V C		HARDCORE FILLING.
151	JOB.	

HARDCORE FILLING

Hardcore filling is the compacted sub-base of floors; it consists of stones, broken sandcrete blocks or coarse gravel. It fills up the space between the subsoil and the soffit of the concrete floor.

FUNCTION OF THE HARDCORE FILLING

The hardcore filling has to carry most of the mass of the concrete floor, except for a small portion supported by the projecting inside edges of the footings. The filling must be well compacted to be firm enough to withstand the weight of the floors.

In addition, the hardcore filling must be built up in such a way that it prevents moisture from rising through it to penetrate the concrete floor.

METHODS OF FILLING AND COMPACTION

The topsoil is removed first (Figs. 1 & 2), then the hardcore filling is added in layers no more than 15 cm deep. Each layer is compacted very well before the next is added.

The bottom layer consists of small rocks, stones or broken sandcrete blocks. The second layer consists of smaller stones. Coarse gravel is used for all the remaining layers, up to about 6 cm below the tops of the footings. The last 6 cm or so is filled with fine sand, which seals off the surface so that no cement is wasted during the floor construction.

As can be seen in Fig. 3, the structure of the hardcore filling becomes denser and finer with each layer, starting from the bottom to the top. This is the correct way to protect the floor from the penetration of moisture.

All layers are compacted with rammers. The coarse gravel can be watered down to ensure proper compaction and to ease the work. If there is a tractor available, it may be used to move the fillings and to speed up the heavy work.

NOTES:

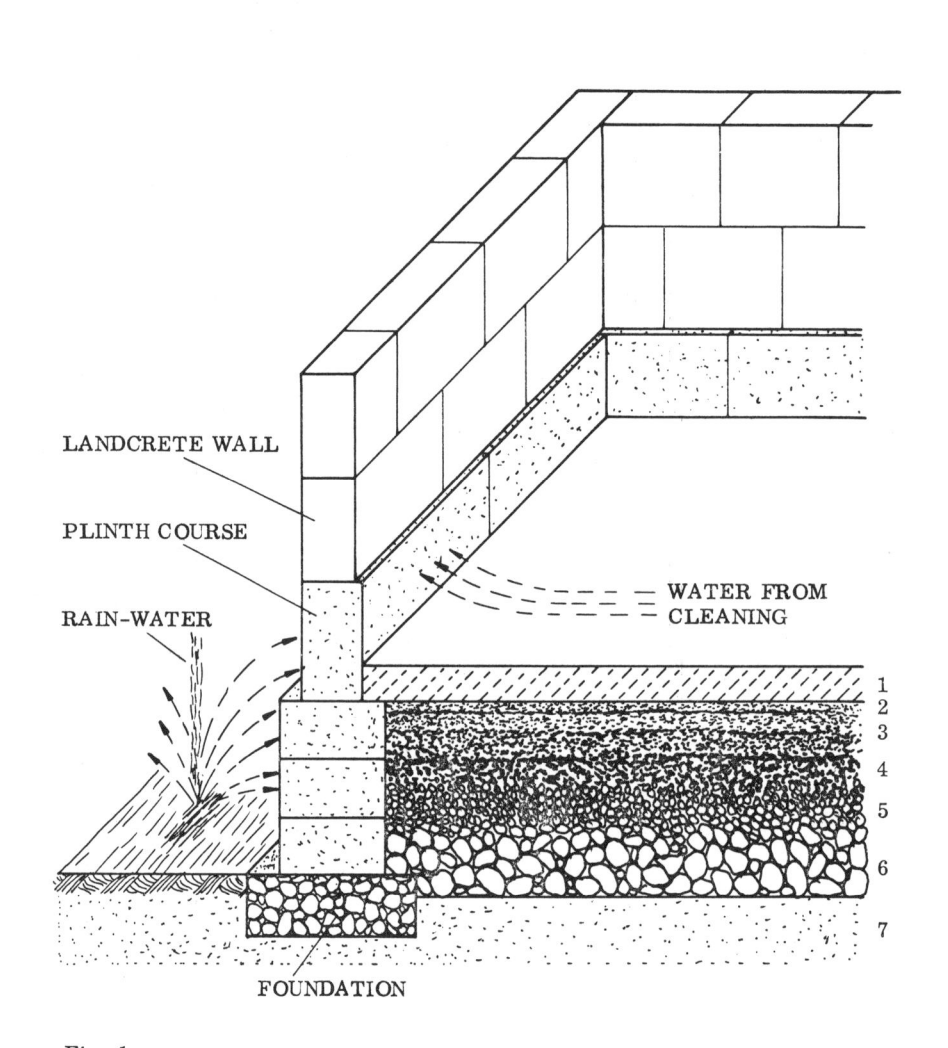

Fig. 1

1 - CONCRETE FLOOR
2 - FINE SAND
3 - FINE GRAVEL
4 - COARSE GRAVEL
5 - SMALL STONES
6 - ROCKS & STONES
7 - FIRM SOIL

N P V C		
153	JOB.	PLINTH COURSE.

PLINTH COURSE

The plinth is a slightly thicker course at the base of a wall or a column; often made of a more durable material than the rest of the wall or column.

In Rural Building, the plinth commonly consists of only one sandcrete course. The plinth course forms the first course of the rising wall immediately above the footings, and it is 1 cm wider than the landcrete blocks (Fig. 1). This 1 cm difference is evident from the inside face of the wall, but it is covered when the wall is plastered.

FUNCTION OF THE PLINTH COURSE

The plinth course raises the landcrete blocks above the finished floor level so that they cannot be penetrated by moisture (from outside by rain, from inside by water used for cleaning, etc.).

Although the landcrete blocks can withstand a skin-deep penetration of water for a short time, they must be protected against the long-lasting influences of the rainy season.

The most affected part of a building is always at the foot of the walls. Rain-water coming from the roof splashes up against the wall and creates a dirty strip about 60 cm in height, which is seen all along the footings. This area is more exposed to penetration by water than the rest of the wall, but the landcrete blocks are raised by the plinth course well above the endangered zone.

The illustration on the opposite page shows the possible paths which the water can take when penetrating the structure.

NOTES:

CORNER BLOCKS

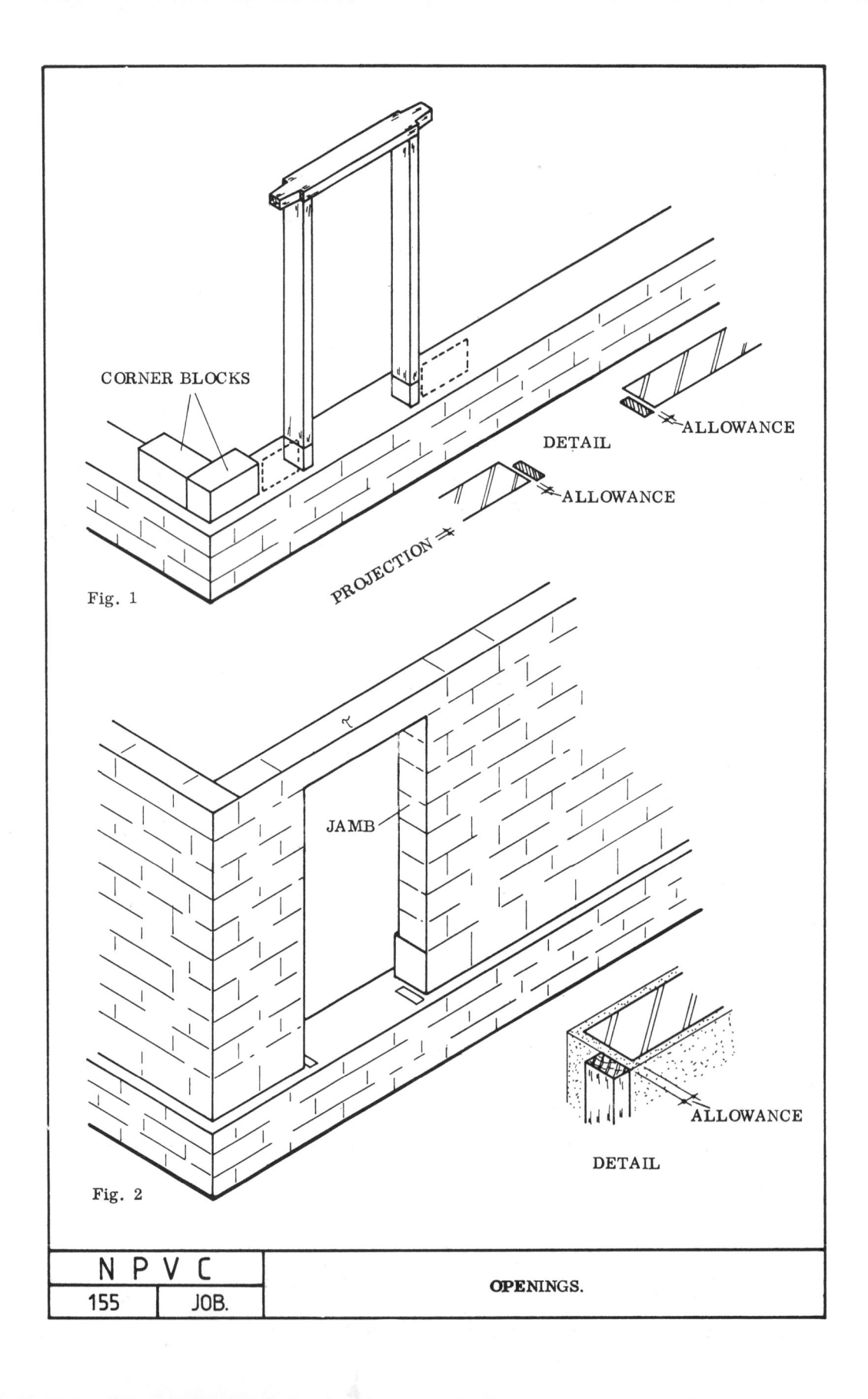

ALLOWANCE

DETAIL

ALLOWANCE

PROJECTION

Fig. 1

JAMB

ALLOWANCE

DETAIL

Fig. 2

N P V C		
155	JOB.	**OPENINGS.**

OPENINGS

An opening is a space in a wall left open for a door or a window. The first openings to be made in walls are the door openings.

DOOR OPENINGS

Before the plinth course is built, the door openings are marked in their correct positions on the footings.

In Rural Building, the door frames are usually made and set into place on the footings before the plinth course is built against the frames (Fig. 1).

If for some reason the door frames are to be installed later, the jambs of the openings (Fig. 2) are built up as described in the chapter on stopped ends, page 22.

A jamb is the portion of wall, or wall face, at the side of an opening. The jambs are built a little wider than the outside measurement of the frame (Fig. 2) so that the frame can fit into the wall opening.

Walling then continues until the window cill level is reached. This is the height where the window openings are set.

WINDOW OPENINGS

The window frames are set and braced before the walling between them is completed. If the frames are to be set later, the window openings must also be built a little wider than the frames, so that the frames can be fitted in later.

In case the opening is to be filled with a decorative grille (Reference Book, page 193) or ventilation blocks (Reference Book, page 195), it is advisable to complete course by course including the special blocks.

This is done because it is easier than putting the decorative blocks into an opening later, and it provides the openwork screen blocks with more stability within the wall during construction.

NOTES:

Fig. 1

CORRECT
- THE TOP SET OF BLOCKS IS SET CROSSWISE TO THE BOARDS

WRONG
- BOARDS PROJECT TOO FAR
- TOP BLOCKS ARE LAID LENGTHWISE TO THE BOARDS AND ARE LESS STABLE

Fig. 2

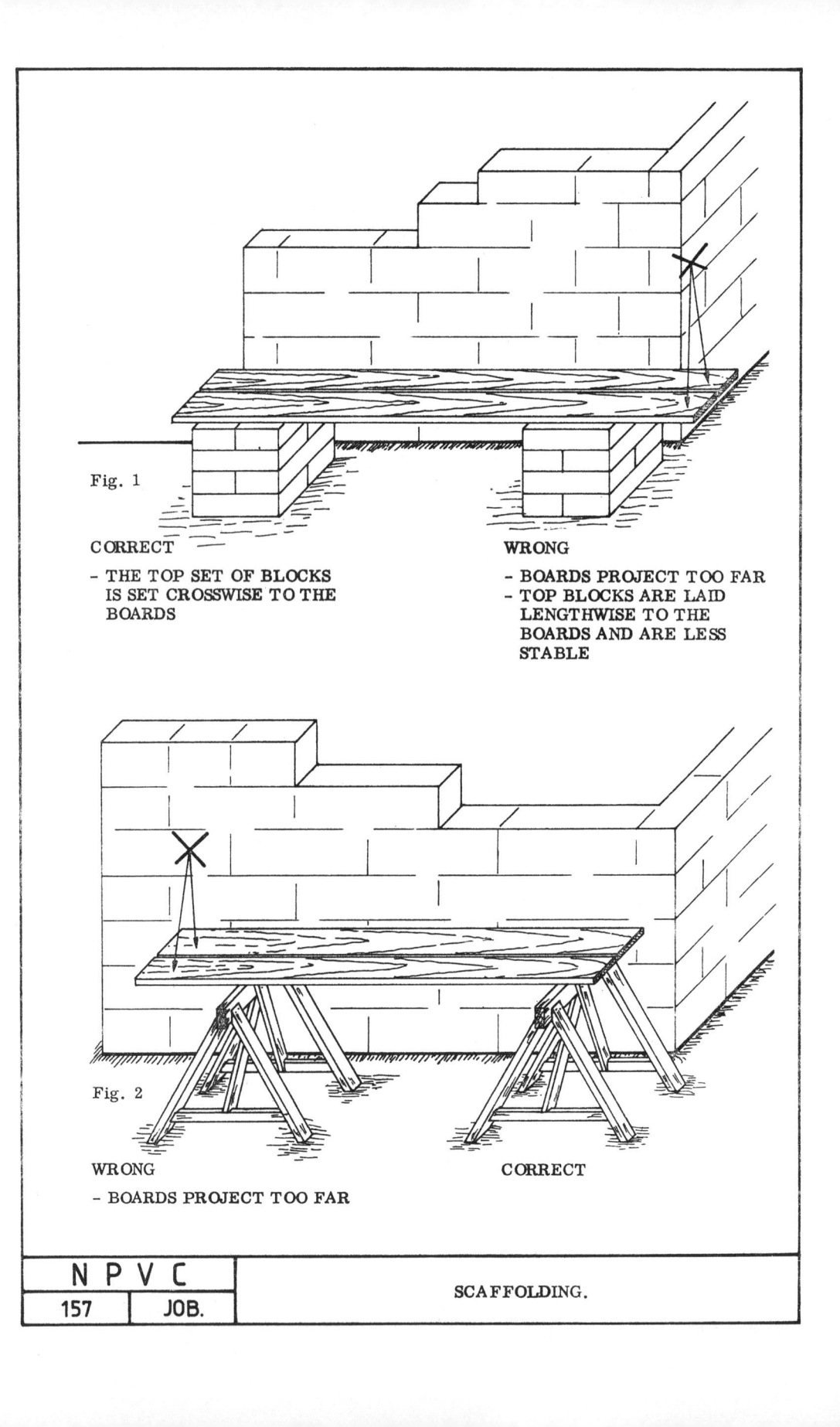

WRONG
- BOARDS PROJECT TOO FAR

CORRECT

N P V C	
157	JOB.

SCAFFOLDING.

SCAFFOLDING

A scaffold is a temporary structure which supports workers and materials during building and other work. It can be made of steel, aluminium, timber or bamboo.

According to their functions, there are three main types of scaffolds:

- Working scaffolds
- Protecting scaffolds
- Supporting scaffolds

Each type may be erected separately and serve only one purpose. Some situations however, require a combination of two or even all three types.

WORKING SCAFFOLDS

As the name indicates, the working scaffold is used for working from. It holds the worker at a height which enables him to comfortably complete walls etc. when the construction has proceeded to a level that makes it difficult to work from the ground. The following is a description of two simple scaffolds used to complete walling between door and window frames. Protecting scaffolds, supporting scaffolds and a number of more complicated scaffolds and their construction will be treated in the Construction Books.

- BLOCK SCAFFOLD: This is the lowest and simplest working scaffold. It is used to raise the worker a bit higher to make it easier to build the wall up to the actual scaffold height of 1,5 m. Set the sandcrete blocks on solid, level ground and lay one or two boards across them (Fig. 1).

- TRESTLE SCAFFOLD: This is a low-level scaffold, consisting of wooden trestles covered by two or three boards (Fig. 2). A trestle is a horizontal beam of wood with two legs on each end. Two or more of them are used to support the boards. Set them on firm, level ground, no further than 1,5 m apart. The height of the trestles can be from 75 to 100 cm. Small quantities of blocks and mortar can be kept on the platform. This scaffold enables the worker to continue walling up to lintel height, and to erect the formwork of the lintels.

NOTES:

Fig. 1

Fig. 2
ENLARGED
DETAIL

NOTCH

N P V C		SCAFFOLDING.
159	JOB.	

LADDERS

A ladder consists of two lengths of wood, metal or rope, called rails; which are connected at a certain distance from each other by rungs (Fig. 1).

Ladders are used to climb and descend scaffolds, walls, etc. during the construction of a building, and to do light maintenance work from later. Since such a piece of equipment is needed not only during the construction but also later in and around the house, it should be made from sound timber in a proper way so it can be used as a permanent ladder.

- CONSTRUCTING THE LADDER: Both the rails and rungs are made out of Odum. The dimensions of the two rails are 5 x 7,5 x 325 cm, while the ten rungs are 2,5 x 5 x 75 cm. The wood should be straight grained and planed to the above sizes.

  Round the top ends of the rails where they touch the wall and cut off the foot ends as shown in Fig. 1, with a 45 degree bevel from each side. To prevent splinters and help the hands to move safely up and down the rails, round off all the edges. Incidentally, when climbing or descending ladders, keep your hands off the rungs. Grip the rear edge of the rails.

  The over-all width of 75 cm gives an inside width of 65 cm which is wide enough to hold two scaffold boards on the rung. Later we will discuss making a ladder scaffold, in the Construction Book.

  The distance of 30 cm between the rungs must not be exceeded, as a wider spacing would make it difficult to climb the ladder.

  The rungs are inserted in notches cut in the rails, and fastened by using 75 cm nails as shown in Fig. 2.

  Sometimes ladders are reinforced by fastening iron rods, threaded at both ends, through both rails behind the first, centre, and last rungs. The rods are held in place by nuts on the threaded ends.

- NOTE: A good ladder should be treated with oil or another preservative. Dried-out wood is the greatest threat to the safeness of the ladder. Always keep ladders in the shade when they are not in use.

NOTES:

Fig. 1

Fig. 2

Fig. 3

Fig. 4

Fig. 5

Fig. 6

N P V C

161   JOB.

BRIDGING OPENINGS.

BRIDGING OPENINGS

To bridge something means to connect the two sides or parts. This can be done in building with a structure of wood, stone, block work, steel or concrete. The essential thing is that the door and window openings must be safely bridged so that the walls above or other members of the structure cannot collapse and damage the house or the people inside.

METHODS OF BRIDGING

There are various ways of bridging an opening. Which one of these is used depends on the distance to be bridged, the shape of the opening and the materials available.

One method which was common in former times was to make the openings so small that a single stone could be laid across them (Fig. 1). Openings which are low enough can also be bridged by inserting supporting blocks arranged like a "V" (Figs. 2, 3, & 4). This method is still common in the dry areas of Africa.

The wider an opening becomes, the more difficult it is to bridge. Builders from many parts of the world eventually learned that if the blockwork remains closed above an opening, both sides of the wall will support each other, maintaining stability. This knowledge was often applied not only to bridge openings but also to construct roofs. The technique was to let each course overlap the one below until the blockwork met at the top and the two sides bonded into each other (Fig. 5).

From this simple method, arches were the next development. They are much stronger and have a more attractive appearance (Fig. 6).

Lintels were introduced in order to produce a square opening into which a frame could be fitted.

NOTES:

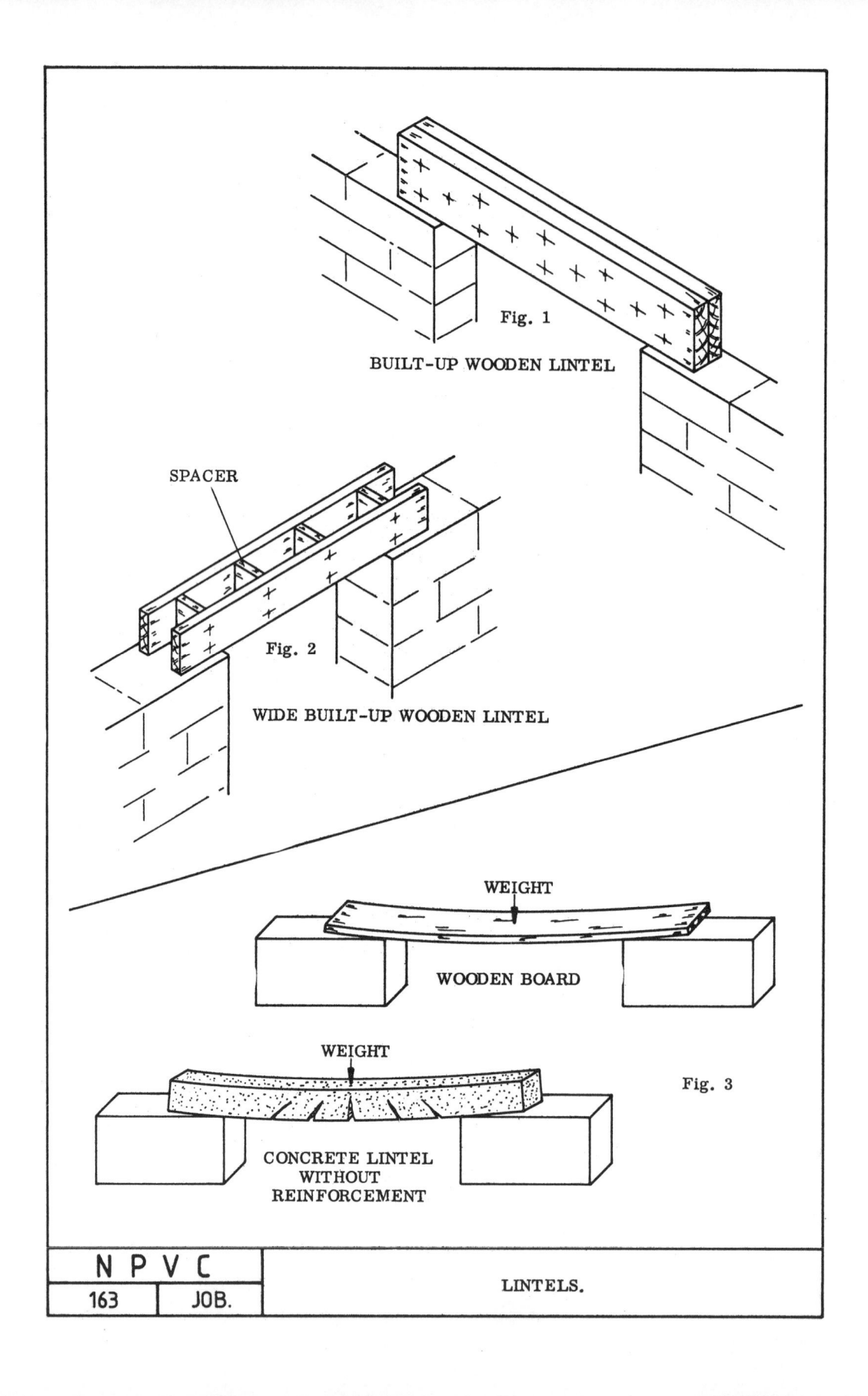

Fig. 1
BUILT-UP WOODEN LINTEL

SPACER
Fig. 2
WIDE BUILT-UP WOODEN LINTEL

WEIGHT
WOODEN BOARD

WEIGHT
CONCRETE LINTEL WITHOUT REINFORCEMENT

Fig. 3

N P V C		
163	JOB.	LINTELS.

LINTELS

BUILT-UP WOODEN LINTEL

A lintel, whatever material it is made from, is a horizontal member of the structure which bridges an opening (Fig. 1).

Its function is to distribute the weight of the blockwork above and any other loads to the supporting walls. With a lintel the opening of the door or window can be lower than with an arch, and it is also easier to fit a frame into the opening.

For short-span openings such as doors and smaller windows which have no additional loads above, the wooden built-up lintel can be employed.

This lintel consists of two or more hardwood boards which are nailed, bolted or screwed together (Fig. 1). In order to save materials when a wider lintel is required, the boards are sometimes connected with spacers between them (Fig. 2).

For long-span openings and in situations where additional loads are present, arches or reinforced lintels are used.

REINFORCED CONCRETE LINTEL

Like the human body which is strengthened by bones, reinforced concrete is made stronger by the steel bars or metal netting embedded in it.

It is known that concrete alone can withstand enormous pressures, but if it is exposed to tensile stresses it will break (Reference Book, pages 168 & 169). Fig. 3 shows that a long board supported only on its ends will bend if weight is set on it. Similarly, a pure concrete lintel will try to bend under a heavy weight, but because the concrete is not flexible this will result in cracks forming across the soffit face, or even in collapse (Fig. 3).

To prevent this, reinforcement bars are embedded in the concrete if the lintel is expected to receive tensile stresses. The combination of concrete and steel does the job where one of them alone would not work: the concrete resists all pressure while the embedded steel resists all stresses.

The reinforcement for a lintel consists of several members with different diameters, shapes and functions. The members are often assembled beforehand in the form of so-called cages (see Fig. 1, next page).

NOTES:

CROSS-SECTION

ERECTION BARS

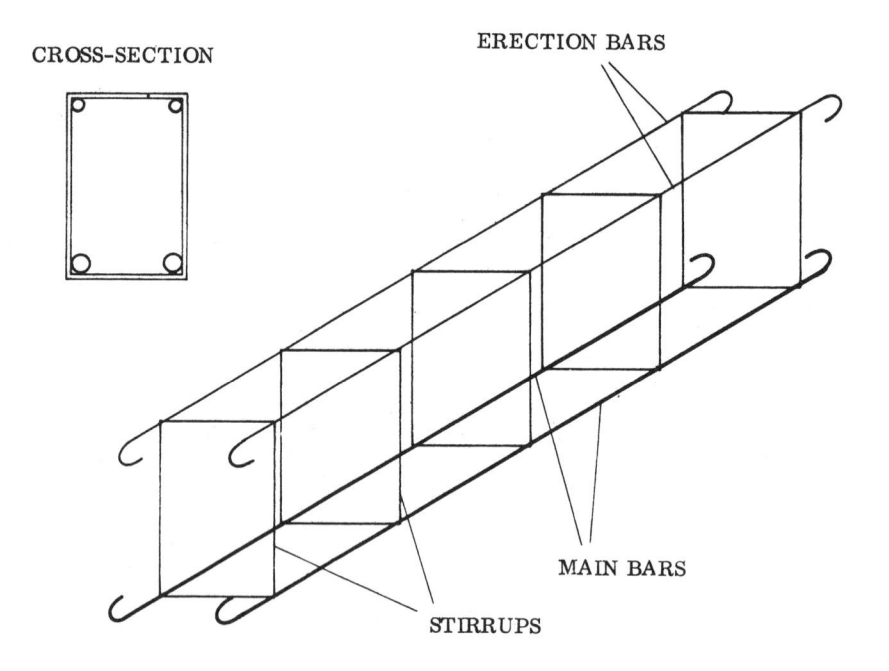

MAIN BARS

STIRRUPS

Fig. 1
REINFORCEMENT CAGE

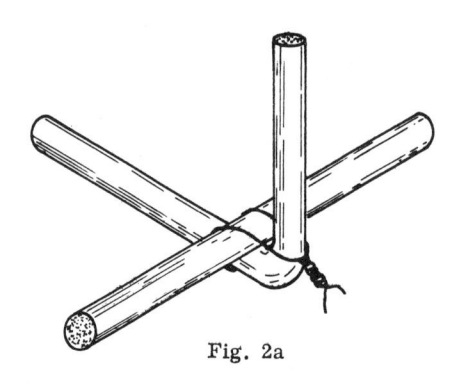

Fig. 2a
STIRRUP TO MAIN BAR

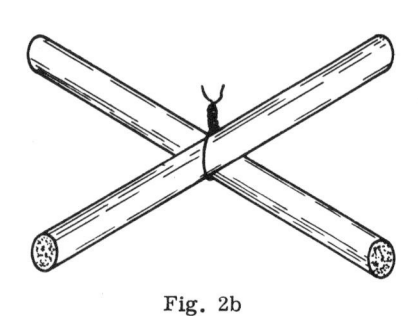

Fig. 2b
BINDING OF CROSSING BARS

N P V C		LINTELS.
165	JOB.	

- REINFORCEMENT CAGE: A typical reinforcement cage is shown in Fig. 1. The lower bars are the main bars. They are normally 12 mm in diameter because they do the actual strengthening of the lintel. The upper bars are thinner and are called erection bars. Since their main function is to hold the cage together, their diameter is only 6 mm. The square-shaped pieces are called the stirrups. They hold the main bars and the erection bars in position. All the different members are bound together with binding wire.

- CUTTING: All reinforcement bars have to be carefully measured and marked off before they are cut. A well equipped building site will have two different sized bolt cutters available. One is to cut the bars ranging from 4 to 10 mm in diameter and one is for bars up to 19 mm. If bolt cutters are not available, a hacksaw or a chisel may be used instead (Reference Book, page 19).

- BENDING: The bars are bent with the aid of the bending plate which is fixed on the work bench; and the bending bars (Reference Book, page 23). Each diameter of bar has to be bent with a specific size of bending bar.

  The bending is done according to a certain radius (this refers to the sharpness of the bend) in order to prevent overstraining and cracking of the bar (this will be explained in more detail later). The stirrups for the cages may be bent around a peg that has the same diameter as the main bars.

- BINDING: The members must be bound together in order to ensure that the bars remain in the correct positions while the concrete is being poured. Binding wire is bound around pieces at the connections (Figs. 2a & 2b). The wire should be stretched taut and then twisted tight with the pincers.

- NOTE: Examine the bars before you use them to make sure that they are free from paint, grease, loose scale or mud. Slight rusting will do no harm, but any loose rust should be removed.

For additional information on concrete and reinforcement steel, see the Reference Book, Materials and Products sections.

NOTES:

LINTELS.

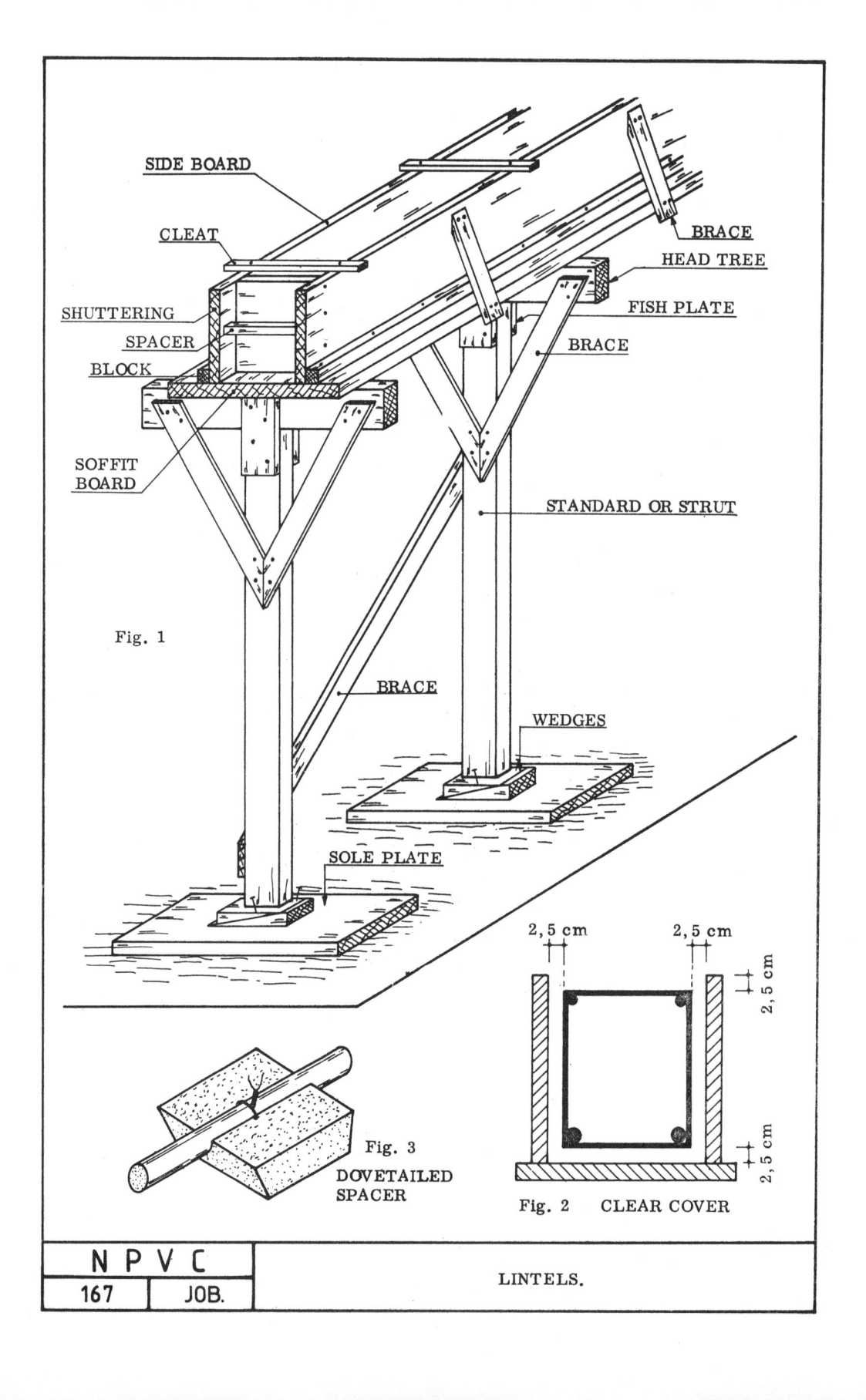

SIDE BOARD
CLEAT
SHUTTERING
SPACER
BLOCK
SOFFIT BOARD
BRACE
HEAD TREE
FISH PLATE
BRACE
STANDARD OR STRUT
BRACE
WEDGES
SOLE PLATE

Fig. 1

2,5 cm   2,5 cm
2,5 cm
2,5 cm

Fig. 3 DOVETAILED SPACER

Fig. 2 CLEAR COVER

N P V C	
167	JOB.

LINTELS.

FORMWORK FOR A REINFORCED CONCRETE LINTEL

The form is simply a temporary box into which the freshly mixed concrete is cast and kept until it has hardened. The inside shape of the box will be the outside shape of the concrete member.

All parts and members of a formwork used for casting reinforced concrete lintels are shown in Fig. 1 on the opposite page. The parts are nailed together in such a way that they can easily be taken apart after the concrete has hardened.

Formwork is made of wood or metal and consists of two structural parts: the shuttering and the strutting.

- SHUTTERING: The shuttering is the actual shaping part of the formwork which is in direct contact with the concrete. Usually Wawa boards are used for shuttering because they are soft and light-weight, thus easy to work with.

- STRUTTING: The strutting is the supporting and bracing part of the formwork. It keeps the shuttering in position and supports both the shuttering and the concrete inside it until the concrete has set hard. Odum boards are usually used for the strutting because they are harder and stronger than the Wawa.

- CONCRETE COVER: Concrete cover, also called clear cover, is the thickness of concrete between the surface of the concrete and the nearest reinforcement bar enclosed in the concrete (Fig. 2).

When the formwork is ready, the reinforcement cage is set into it. Spacers are attached to the bottom side of the stirrups (Fig. 3). These ensure that the reinforcement bars are correctly positioned within the concrete. The spacers are made beforehand out of cement mortar to the dimensions specified for the concrete cover thickness. A short piece of binding wire should be pressed into the fresh mortar of the spacer, so that it can be fixed properly on the rod.

There must also be spacers on the sides of the cage, to hold the stirrups away from the side boards of the shuttering.

- NOTE: When casting concrete, take care that the cage remains in position and that the concrete is well compacted around the reinforcement bars.

NOTES:

LINTELS.	N P V C
	JOB. 168

CROSS-SECTION

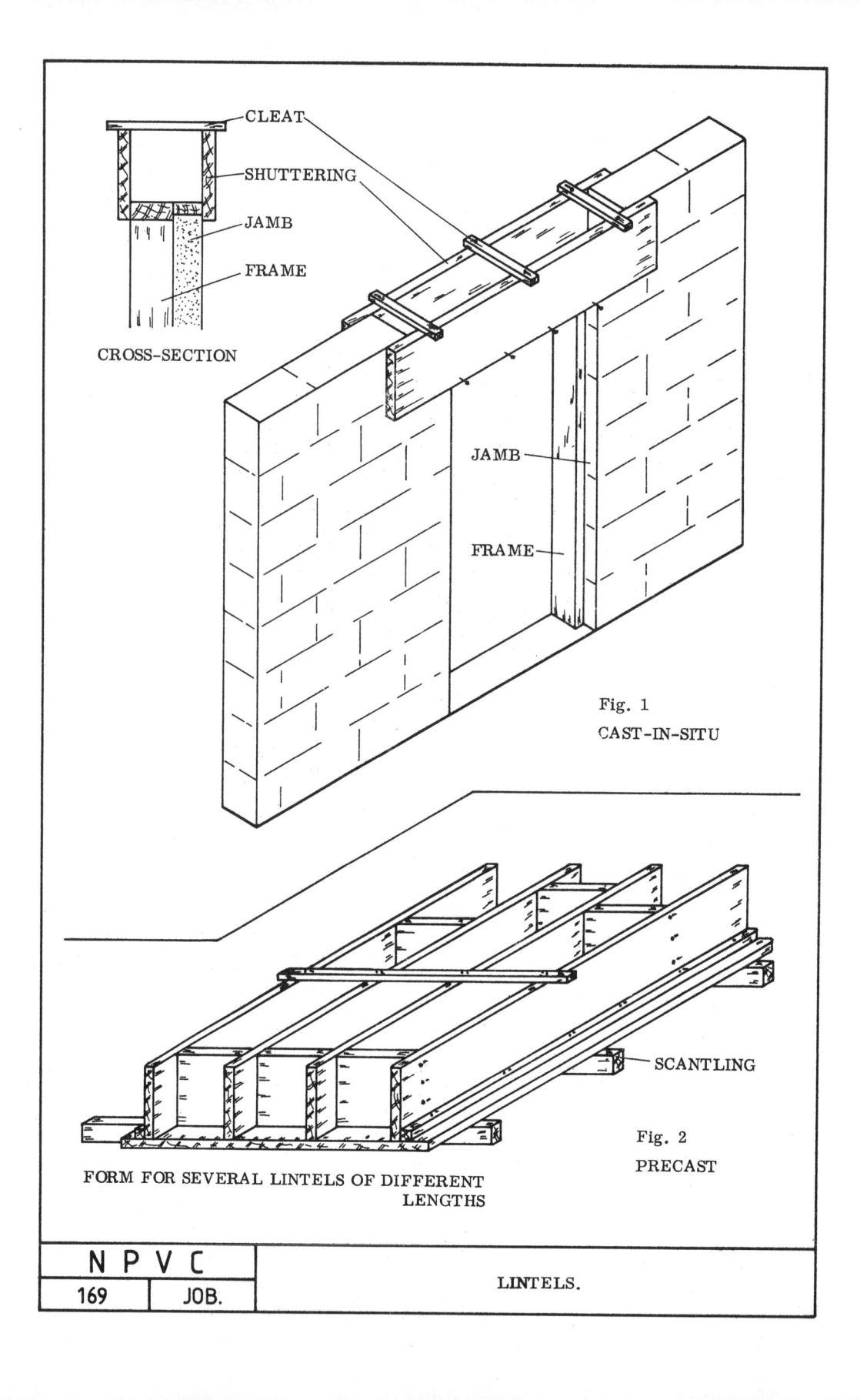

- CLEAT
- SHUTTERING
- JAMB
- FRAME

JAMB

FRAME

Fig. 1
CAST-IN-SITU

SCANTLING

FORM FOR SEVERAL LINTELS OF DIFFERENT LENGTHS

Fig. 2
PRECAST

N P V C		
169	JOB.	LINTELS.

CASTING REINFORCED CONCRETE LINTELS

Concrete lintels are made in two ways. They are either cast-in-situ or they are precast.

- CAST-IN-SITU: This method is the most common one in Rural Building. The lintel is cast in situation; in the place where it is needed (Fig. 1). The advantage of this method is that no soffit board is needed for the form, because the head of the door frame acts as the bottom of the form, provided that the frame has been installed already. A further advantage is that any roof anchorage, if needed here, can easily be inserted into the reinforcement cage at the correct position.

- PRECAST: A precast concrete lintel is a lintel which is made in advance. The formwork is on the ground and the concrete is cast there. When the construction reaches lintel height, the concrete lintel has set hard and can be set into position. The advantage of this method is that it saves time, since the wall above can be completed immediately after the lintel has been laid (Reference Book, page 137).

  If several lintels have to be made, a form like the one shown in Fig. 2 can be used.

- NOTE: Precast concrete lintels must be marked on their top face with the letter "T" to ensure that they are placed in the correct position and not upside-down.

- PREPARATION OF THE FORMWORK: No matter which of the above methods is used, the formwork has to be prepared before the concrete is cast. This involves cleaning dirt and dust from the surfaces which will be in contact with the concrete, and watering or oiling them.

  The formwork must be completely sealed so that no gaps remain for the cement paste to escape. This would result in voids and weak concrete.

- COMPACTION: The concrete is filled into the formwork in layers, and compacted by tamping with an iron rod or the trowel. Tapping lightly on the formwork with a hammer also helps to consolidate the concrete.

- CURING: When the concrete is starting to get hard, the lintel must be kept wet and covered. This process is called curing and must be continued until the concrete is completely set and the formwork can be removed.

NOTES:

LINTELS.	N P V C
	JOB. 170

- STRIPPING: Stripping refers to the removal of the formwork; this has to be done carefully to avoid causing shocks or vibrations.

    After the formwork is removed, clean all the parts of it and remove the nails. Stack the different parts neatly to keep them from getting bent or warped.

- STRIPPING TIME: This is the period between the casting of the concrete and the time the formwork can be stripped off. During this time the formwork containing the fresh concrete must not be disturbed, so that the concrete can set hard without any cracks forming in it.

    Depending on the size, shape and position of the concrete member, the stripping time varies from 4 to 28 days.

NOTES:

ROOFS

ANCHOR BEAM

When the walling above the lintels is completed, the top of the wall may be covered with a strip of reinforced concrete, all around the outside walls of the house. This is the anchor beam, where the roof construction is anchored. It is also known as the ring beam or concrete belt.

WALL PLATE

For smaller spans, the Rural Builder can install a wooden wall plate instead of an anchor beam. Both of these are explained further in the section on roofs in the Construction Book.

TERMS

On this page and the next page, most of the parts and members of the roof construction are mentioned. The details of roof construction are dealt with in the Construction Book.

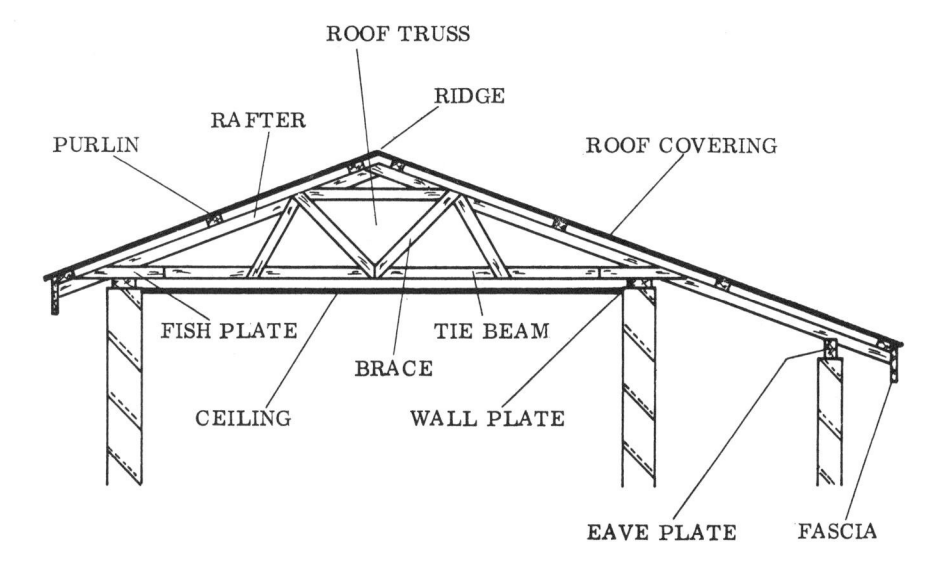

ROOFS.

NPVC JOB. 172

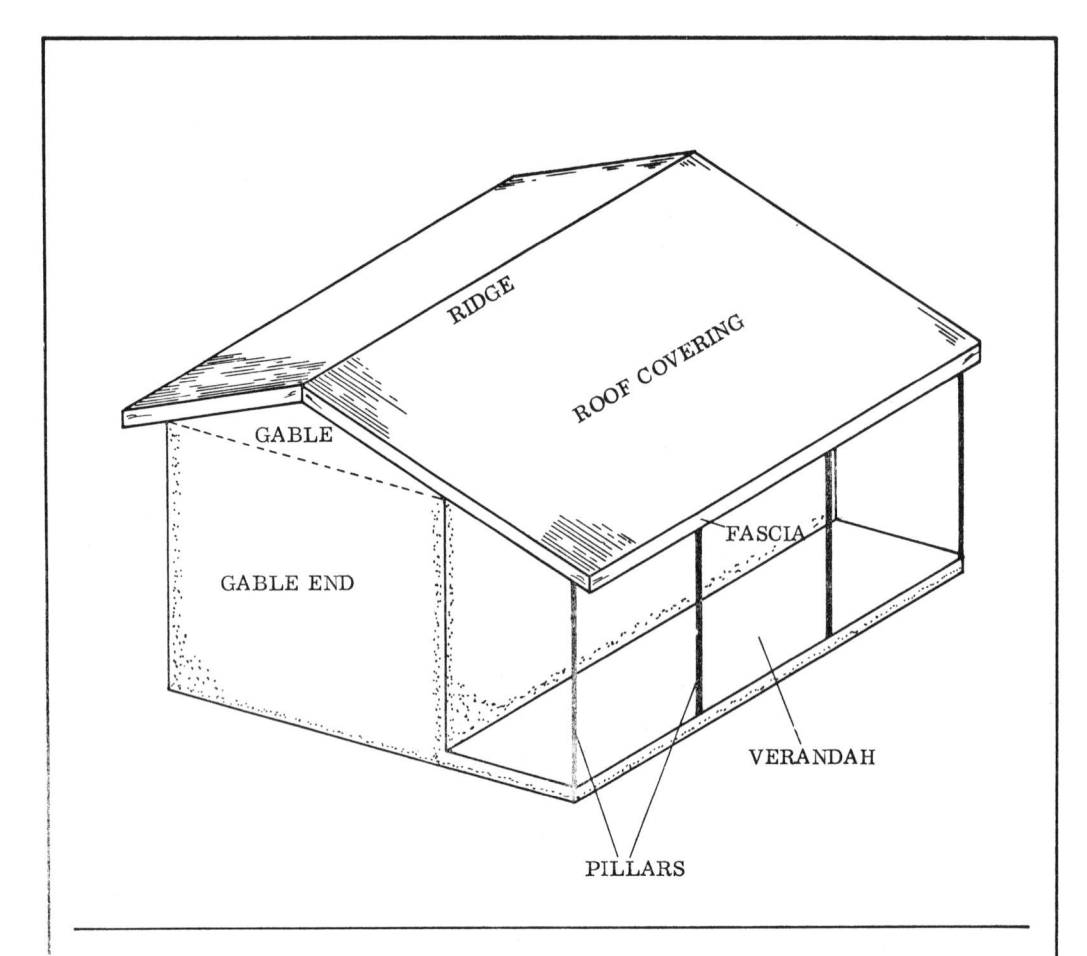

Labels on diagram: RIDGE, ROOF COVERING, GABLE, GABLE END, FASCIA, VERANDAH, PILLARS

NOTES:

N P V C

173 | JOB.

ROOFS.

PLASTER - RENDER

Plaster and render are mortars with different mix proportions; they are applied to walls to protect the blocks from weather etc.

Internal surfaces are plastered; the mortar used is called plaster.

External surfaces are rendered; the mortar is called render.

FUNCTIONS OF PLASTER AND RENDER

Since most of the wall consists of landcrete blocks, it has to be rendered to make it weatherproof, and to protect the blocks against rain which otherwise quickly damages them. The ideal rendering will prevent water from penetrating, will be free from cracks and will stick tightly to the wall. At the same time, the appearance of the building is improved.

The function of plaster is to give the inside walls a smooth, plain finish so that the rooms both look nice and are easily cleaned. No gaps or holes should remain for insects, spiders, etc. to find shelter in. Wet areas such as bathrooms, showers, kitchens and toilets, are plastered to protect the landcrete blocks from moisture penetration. In addition, the plaster serves as a protection against fire in buildings which are made out of a wood skeleton covered with mud (a common construction in southern Ghana).

APPLICATION

The application of the plaster can be done with one coat or two coats. Generally one coat work is done, but two coats may sometimes be required, for example when the wall is very uneven and a thick plaster coat is needed to cover the irregularities.

Sometimes a spatterdash coat is applied to the wall before the plaster, to give a good grip to the plaster. This is discussed on page 176; see also the Reference Book, page 29.

NOTES:

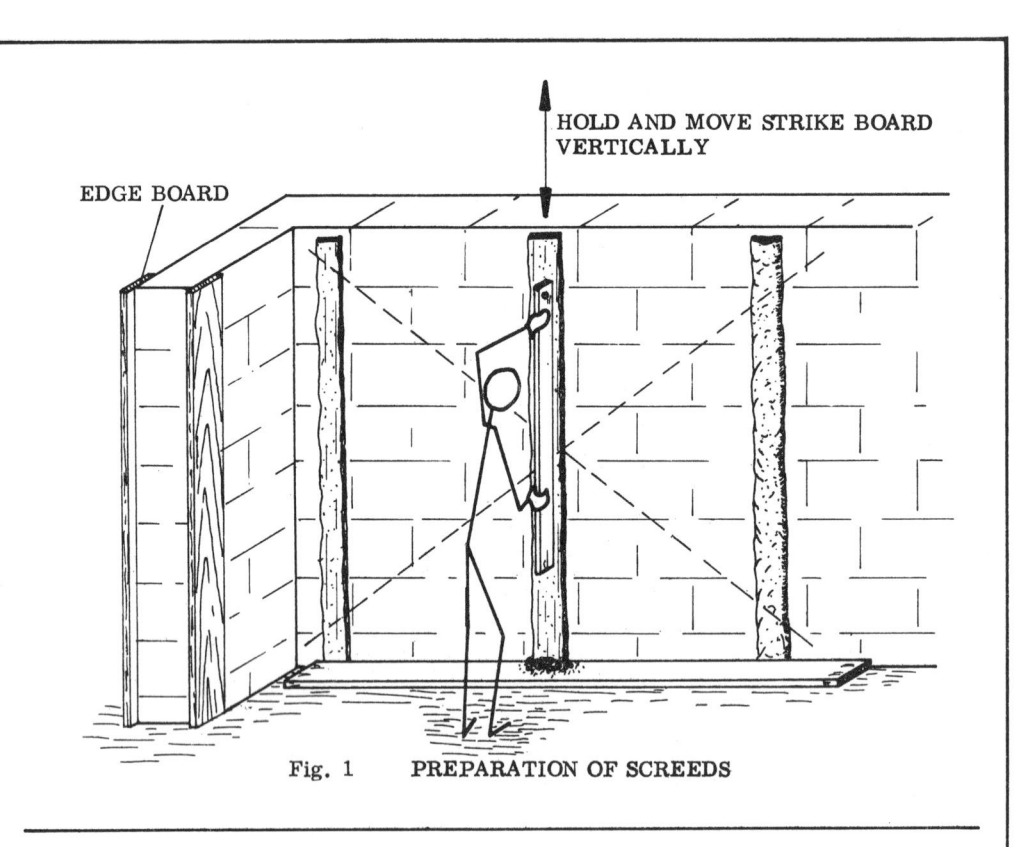

Fig. 1 PREPARATION OF SCREEDS

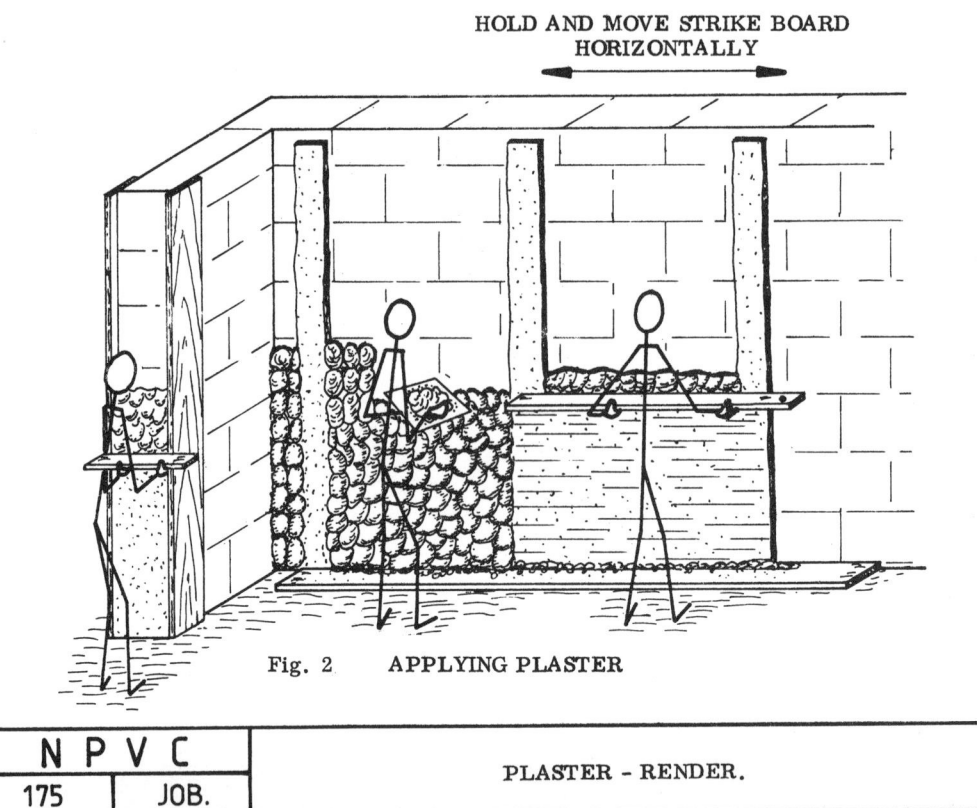

- EDGE BOARD
- HOLD AND MOVE STRIKE BOARD VERTICALLY

Fig. 2 APPLYING PLASTER

- HOLD AND MOVE STRIKE BOARD HORIZONTALLY

NPVC		PLASTER - RENDER.
175	JOB.	

PLASTERING OR RENDERING

Before the plaster or render is applied, the wall should be thoroughly checked to make sure that it is plumb and its surface flat. Holes and hollow parts should be filled in, and single projecting blocks must be chiselled off. The latter problem can often occur on the inside of walls due to irregularly made blocks.

The face of the wall should be free from loose dirt and dust, and it must be well dampened to reduce the absorption of moisture from the mortar.

On outside corners, so-called edge boards are fixed so that they project past the edge by a distance equal to the thickness of the coat to be applied (Fig. 1).

On inside walls and between edge boards, screeds are prepared according to the required thickness; they are situated as wide apart as the strike board (Reference Book, page 25) can readily bridge.

A screed is a strip of plaster or render which is carefully applied to the correct thickness to act as a guide for the strike board. In making the screeds, hold the strike board vertically and give it an up and down motion as you move it across, smoothing the screed (Fig. 1). The screeds will be flush with the finished surface and must be carefully plumbed and lined out (Fig. 1).

A board can be laid close along the bottom of the wall so that any mortar which is dropped can be picked up again, to avoid wasting material.

When the screeds are ready, the plaster can be applied as shown in Fig. 2. For this step the strike board is held horizontally and moved from side to side as it is pushed up the wall, smoothing the plaster.

THE AGGREGATES

Regardless of where the sand for the plaster or render comes from, it must be clean and suitably graded (Reference Book, page 159, and pages 147 to 151).

Sand for rendering should be sharp and well graded, from fine to very coarse. The mortar may be difficult to apply, but nevertheless this is the best way to get a rendering that will not crack and let water in.

Sand for plastering should be well graded, from fine to fairly coarse; and for a finishing coat from fine to medium. Well graded sand reduces the drying shrinkage and cuts down the danger of cracking and crazing. Crazing means the formation of hairline cracks on the surface of concrete, plaster or render; usually it is caused by too much water in the mix, or by a too rich mix.

PLASTER - RENDER. NPVC JOB. 176

MIX PROPORTIONS

In Rural Building, most of the mortars used for plaster and render are cement mortars, because lime is not always available.

The mix proportion for plaster ranges from 1:8 to 1:12, depending on the condition and grading of the sand. Both coats should have the same strength as far as possible if two coats are applied.

The mix proportion for render can vary from 1:6 to 1:10. It is generally agreed that the sides of the house which are exposed to rain should have a better mix. However, the mix proportion should never be better than 1:6, and should be adapted if possible to the strength of the background (Reference Book, pages 158 to 165).

ADDITIONAL PROTECTIVE MEASURES

When the plaster or render has set hard and has dried, it is advisable to paint the whole surface with emulsion paint (Reference Book, page 201).

If lime is available (Reference Book, page 152 and page 200) the walls are first white-washed to fill up the tiny holes in the surface. After this the emulsion paint is applied in two coats. The paint prevents water from penetrating into the plaster or render; reflects light and thus keeps the walls cooler, and gives the house an attractive appearance.

When emulsion paint has dried, it will not dissolve in water, so it can easily be cleaned with water and a soft brush.

- REMEMBER: Sand for plaster and render should be clean, properly graded and as coarse as is appropriate for the particular application.

  Joints should be carefully raked out while the mortar is still fresh, so that the plaster or render can grip well. Another method is to apply a spatterdash to the wall to give a good grip to the plaster or render.

  No coat should be richer than the coat underneath it. If you have to make two of different strengths, then the undercoat should be stronger than the finishing coat. Try to avoid making two coats of plaster or render, because it is difficult to get a good connection between the two coats. The spatterdash is not a coat, it is a background for the plaster or render.

NOTES:

SPATTERDASH

Spatterdash can be used to produce an attractive appearance, or it can be applied before plastering in order to make a good surface for the plaster or render.

Spatterdash is a wet, rich mix of cement and sand, called a slurry. This sand and cement is mixed to a proportion of 1:1,5 or 1:3. This slurry is thrown hard, or spattered, against the smooth block or concrete surface, and then allowed to harden (Fig. 1).

When you work with this hand operated machine (Reference Book, page 29), do not overload the machine with material. It is better to use small quantities at frequent intervals. All of the slurry must be used within 1 hour of the time it is mixed.

Do not set the flicker bar adjuster beyond the second notch when the machine is new. Only when the bar wears out should it be set to a lower notch (Fig. 2).

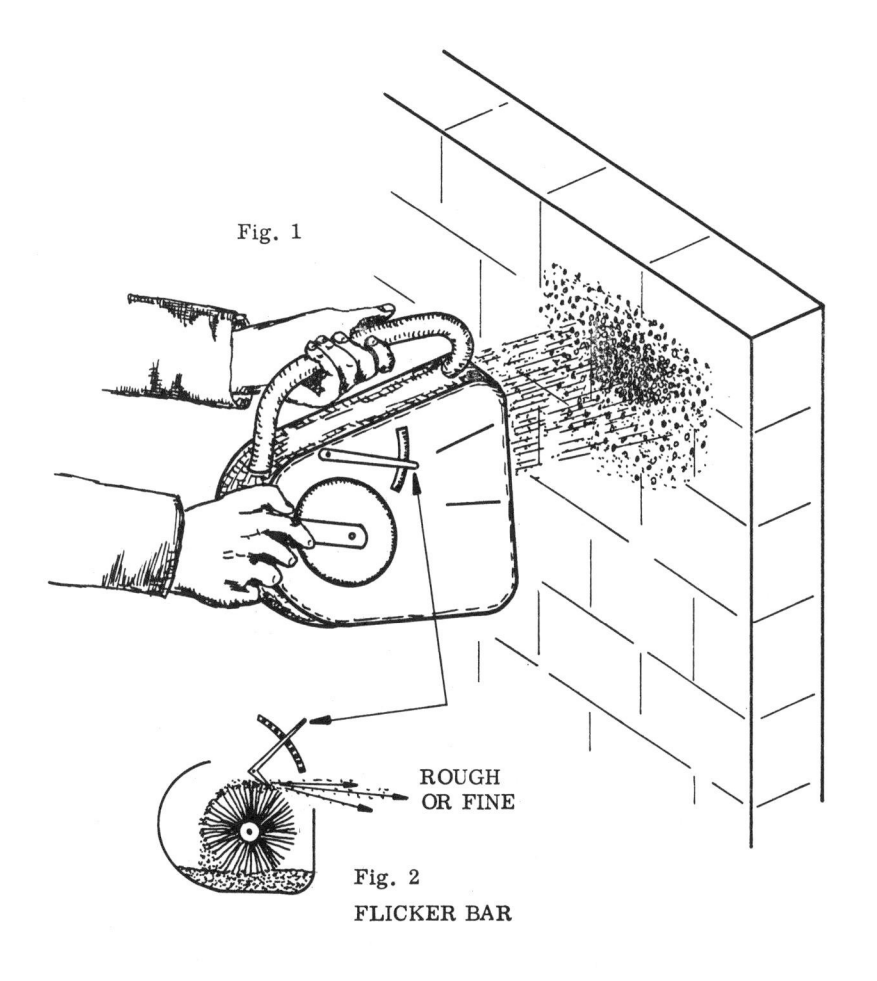

Fig. 1

ROUGH OR FINE

Fig. 2
FLICKER BAR

PLASTER - RENDER.

N P V C

JOB. 178

FLOOR CONSTRUCTION

The dry climate in the north of Ghana makes it advisable to construct the floor after the roof covering has been completed. This makes it easier to cure the concrete, and to make sure that the floor sets hard without cracking due to excessive drying. There are various methods to construct floors. In Rural Building the main ones are: one-course work, and two-course work.

ONE - COURSE WORK

This means that the final surface finish is completed before the base layer has set hard. The result is a monolithic floor construction; which means that the floor throughout can be considered as one solid mass (Fig. 1). The advantage of this method is the short construction time, using a minimum of materials, and no separation between the top layer (screed layer) and the base layer.

TWO - COURSE WORK

This means that the base layer and the finish layer are constructed separately. After the base layer has set hard, a floor screed is applied. This is a fine-grained mortar layer, about 2 cm thick, laid to finish the floor surface (Fig. 2).

The advantage of this method is that any faults in the base layer, such as cracks, can be covered. However, it takes longer to construct and requires more cement. Another disadvantage is that it can be difficult to get a good connection between the base layer and the floor screed. The base layer should be carefully treated with cement and water to form a good connection between the floor screed and the base layer.

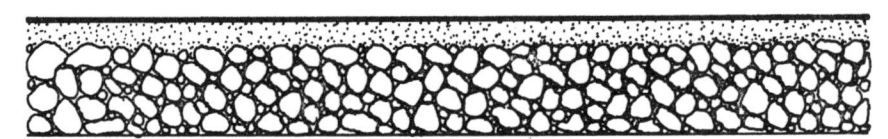

Fig. 1   ONE - COURSE WORK

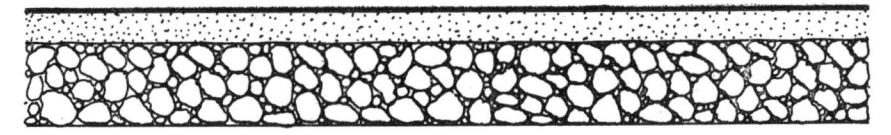

Fig. 2   TWO - COURSE WORK

CASTING METHODS

If a floor area is larger than approximately 10 square meters, the area should be divided into bays for concreting. A bay is one of several uniform divisions of a concrete floor which are cast at any one time.

The bays are separated by edge boards, which are laid and levelled to the required floor thickness. The edge boards act as a guide for the strikeboard to level off the concrete surface, therefore they must be laid and levelled with great care.

On the drawing below you can see the positions of the edge boards when the floor is divided into six bays. The boards should be arranged as shown so that the corners of the concrete bays will match each other when the floor is complete.

The division of the floor into bays helps to prevent the development of cracks due to shrinkage during the hardening process. The smaller the area, the less the shrinkage, and the fewer cracks will appear. Square-shaped bays are the best because all the sides will shrink by the same amount.

Bays also make the construction process easier. The bays are small enough to be cast, levelled and finished within a manageable time. The work can be interrupted to allow the already completed bays to harden. Then the edge boards are removed and the empty bays are cast, using the completed ones as a guide. Once you start casting a bay, it must be completed. Never interrupt a concreting process, as this can result in a faulty bond and the joint will always be visible. For the sequence of operations for casting a floor see Figs. 1 to 8 on the following two pages.

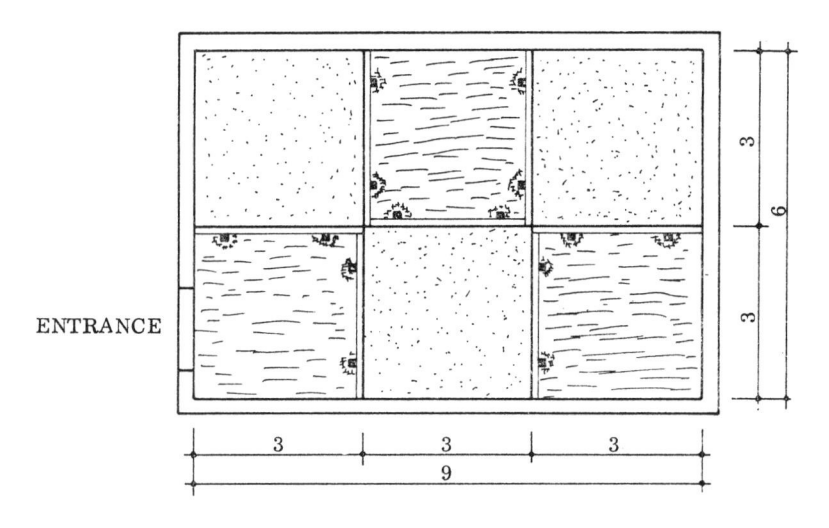

ENTRANCE

DIVISION INTO BAYS

FLOOR CONSTRUCTION.

NPVC
JOB. 180

Fig. 1
- CLEAN THE AREA
- CLEAN THE EDGES OF THE FOOTINGS
- LEVEL THE AREA WITH A SHOVEL

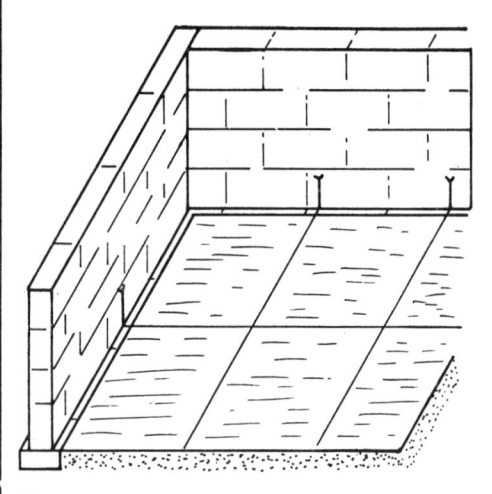

Fig. 2
- MARK THE BAYS ON THE WALLS
- SET OUT THE POSITIONS OF THE EDGE BOARDS ON THE GROUND

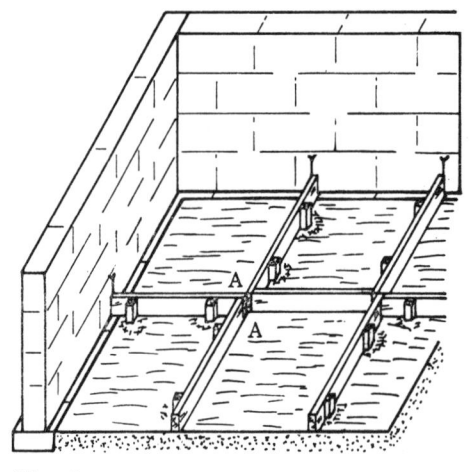

Fig. 3
- FIX THE EDGE BOARDS WITH PEGS
- LEVEL THE TOPS OF THE EDGE BOARDS
- THE CORNERS SHOULD MATCH (A to A)

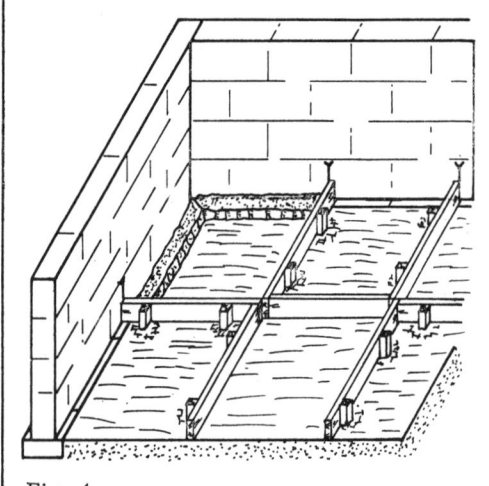

Fig. 4
- POUR GUIDE STRIPS TO THE HEIGHT OF THE BASELAYER
- LEVEL THE GUIDE STRIPS

N P V C	
181	JOB.

FLOOR CONSTRUCTION.

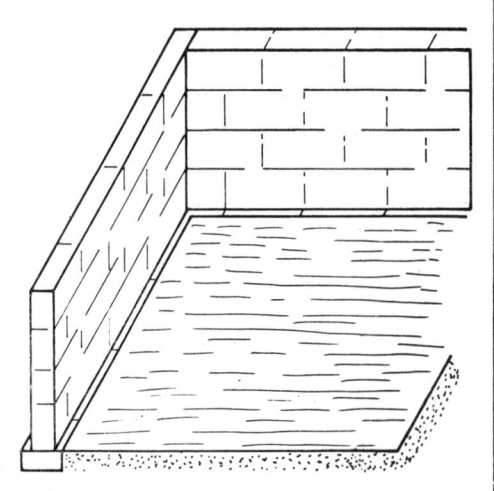

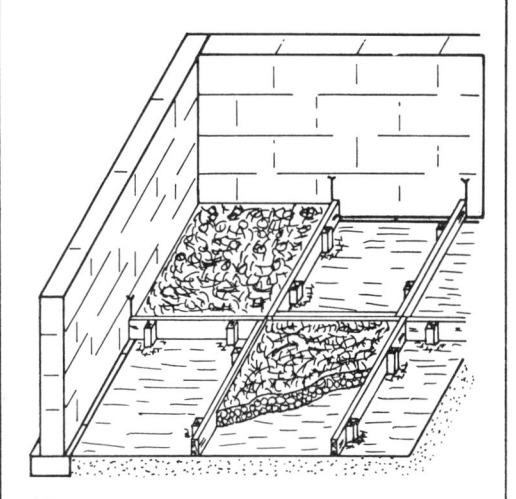

Fig. 5
- POUR THE BASE LAYER AND TAMP IT DOWN WITH THE RAMMER TO THE REQUIRED LEVEL

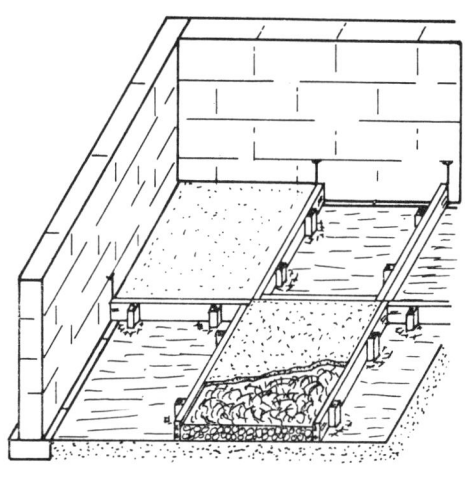

Fig. 6
- LAY THE FLOOR SCREED LAYER (FAIRLY DRY MIX) AND TAMP IT DOWN WITH THE WOOD FLOAT UNTIL MOISTURE COMES THROUGH

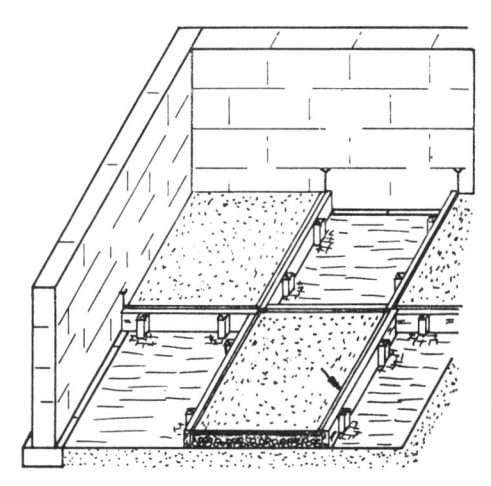

Fig. 7
- FINISH OFF THE TOP WITH THE TROWEL OR STEEL FLOAT
- MAKE A BEVEL ALONG THE EDGES FOR THE SHRINKAGE JOINTS (ARROW)

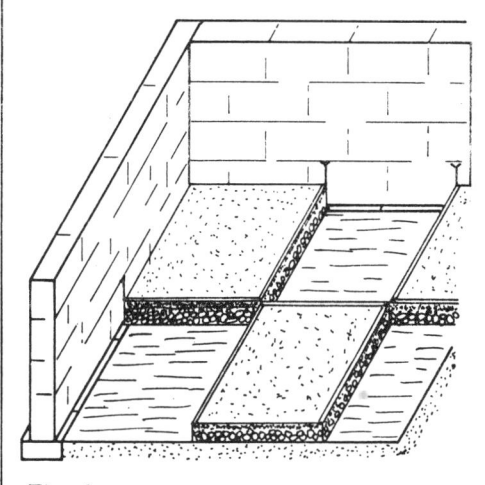

Fig. 8
- TAKE AWAY THE EDGE BOARDS AND PUT SAND ON TOP OF THE FLOOR
- WATER THE FLOOR REGULARLY

FLOOR CONSTRUCTION.

NPVC	
JOB.	182

SHRINKAGE GAPS

When one set of bays has hardened, the edge boards are carefully removed and the remaining bays can be cast. Shrinkage gaps are made between the adjoining bays. This is done by placing plastic or paper between the bays when the second set of bays is cast, so that the bays are kept separate from each other.

The shrinkage gaps allow the concrete bays to shrink a bit as they harden without cracking. This type of gap is used where the floor is not exposed to the sun or to great temperature changes; usually only for inside floors. Shrinkage gaps can be made in either one-course work or two-course work. A "V" is made along the top edges of the gap to improve the appearance of the floor (Fig. 1).

EXPANSION GAPS

Where the floor is exposed to the sun, as in a verandah floor or any concreted area outside the house itself, expansion gaps have to be made. In this case the edge boards are not removed until all the bays have been cast and hardened. The boards are then removed and the gap between the bays is filled with wet sand, and the floor screed is applied over the top (Fig. 2). Expansion gaps can only be made in two-course work.

The expansion gap allows the floor to expand and contract with the temperature changes without forming cracks. Expansion occurs when the floor is heated by the sun during the day, and shrinkage occurs at night when it cools down. Floors exposed to the sun should be divided into bays of no more than 5 square metres, and expansion gaps should be made in them. The floor screed must also be provided with a shrinkage gap; this should be located directly above the expansion gap in the base layer (Fig. 2).

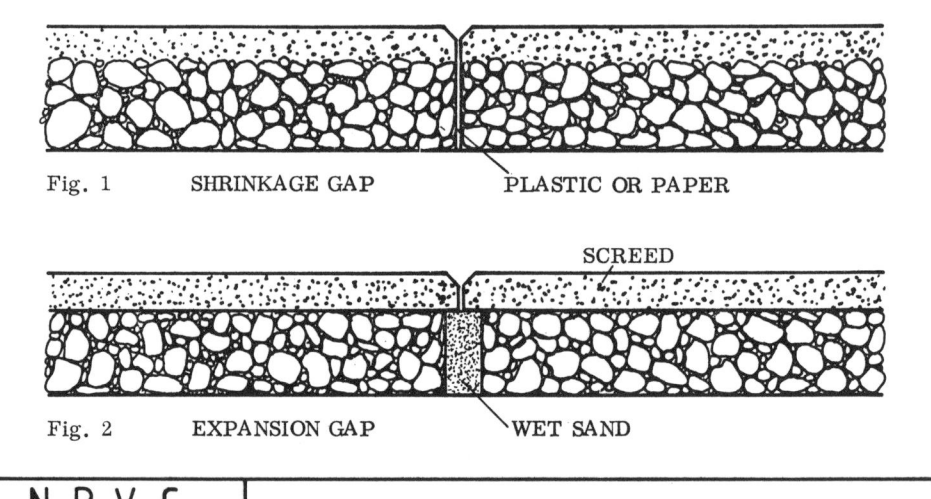

Fig. 1   SHRINKAGE GAP   PLASTIC OR PAPER

SCREED

Fig. 2   EXPANSION GAP   WET SAND

FLOOR CONSTRUCTION.

A WORD FOR THE TRAINEE BEFORE ON-THE-JOB TRAINING

With the basic knowledge you have gained in this part of the course, you should now be prepared to go to a building site for on-the-job training.

Remember that so far you have covered only the first part of the course. You will still need to acquire much more knowledge and many more skills before you can be called a Rural Builder.

While you are working at the building site, remember that this time is also supposed to be a learning experience. If you want to learn, you have to ask questions about how and why certain things are done. If you can't get your questions answered at the time, write them down and bring them to your instructors for explanation. It is a good idea to keep a notebook; writing down in it the methods you applied at the building site, the time that a certain operation required, the materials used and how much was needed, and any ideas you have about how it could have been done differently.

You should occasionally review the information in this book, especially as it comes up on the job. Don't be afraid to use the book and to write notes in it; the notes which you make at this time can be very helpful to you later on, when you have finished the course and are working as a builder.

NPVC